High-Yield

Microbiology and Infectious Diseases

SECOND EDITION

Louise Hawley, Ph.D.
Professor and Chair
Department of Medical Microbiology and Immunology
Ross University School of Medicine
Dominica, West Indies
and
Emeritus Assistant Professor
Department of Medical Microbiology and Immunology
University of Minnesota, School of Medicine—Duluth
Duluth, Minnesota

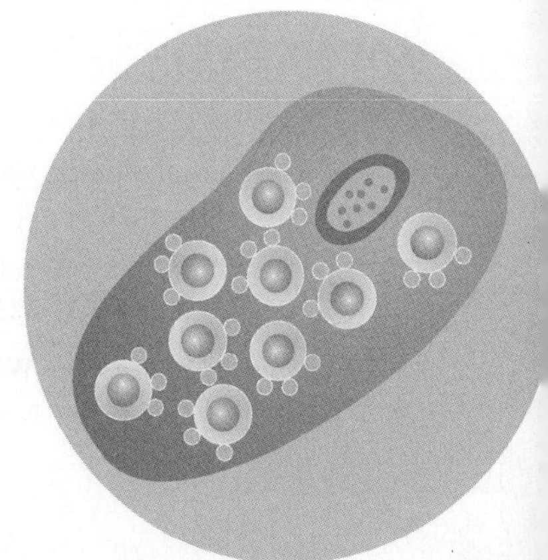

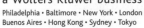

Lippincott Williams & Wilkins
a Wolters Kluwer business
Philadelphia • Baltimore • New York • London
Buenos Aires • Hong Kong • Sydney • Tokyo

Acquisitions Editor: Betty Sun
Managing Editor: Melissa Nolan
Marketing Manager: Emilie Linkins
Associate Production Manager: Kevin P. Johnson
Designer: Teresa Mallon
Compositor: International Typesetting and Composition
Printer: Data Reproductions Company

351 West Camden Street
Baltimore, MD 21201

530 Walnut Street
Philadelphia, PA 19106

Printed in the United States of America

High-Yield Microbiology and Infectious Diseases, Copyright 2000 Lippincott Williams & Wilkins

Library of Congress Cataloging-in-Publication Data

Hawley, Louise
High-yield microbiology and infectious diseases / Louise Hawley. — 2nd ed.
p. ; cm. — (High-yield series)
Includes index.
ISBN 978-0-7817-6032-4
ISBN 0-7817-6032-1
1. Medical microbiology—Outlines, syllabi, etc. 2. Communicable diseases—Outlines, syllabi, etc. I. Title. II. Series. [DNLM: 1. Gram-Negative Bacteria—pathogenicity—Case Reports. 2. Gram-Negative Bacteria—pathogenicity—Examination Questions. 3. Gram-Negative Bacteria—pathogenicity—Outlines. 4. Communicable Diseases—microbiology—Case Reports. 5. Communicable Diseases—microbiology—Examination Questions. 6. Communicable Diseases—microbiology—Outlines. 7. Fungi—pathogenicity—Case Reports. 8. Fungi—pathogenicity—Examination Questions. 9. Fungi—pathogenicity—Outlines. 10. Gram-Positive Bacteria—pathogenicity—Case Reports. 11. Gram-Positive Bacteria—pathogenicity—Examination Questions. 12. Gram-Positive Bacteria—pathogenicity—Outlines. 13. Parasites—pathogenicity—Case Reports. 14. Parasites—pathogenicity—Examination Questions. 15. Parasites—pathogenicity—Outlines. 16. Viruses—pathogenicity—Case Reports. 17. Viruses—pathogenicity—Examination Questions. 18. Viruses—pathogenicity—Outlines. QW 18.2 H396h 2007]

QR46.H37 2007

616.9'041—dc22 2006007932

To purchase additional copies of this book, call our customer service department at **(800) 638-3030** or fax orders to **(301) 223-2320.** International customers should call **(301) 223-2300.**

Visit Lippincott Williams & Wilkins on the Internet: http://www.LWW.com. Lippincott Williams & Wilkins customer service representatives are available from 8:30 am to 6:00 pm, EST.

06 07 08 09 10
2 3 4 5 6 7 8 9 10

To my wonderful, supportive, and tolerant husband and daughters, and to my mother, who gave me many essential early lessons on teaching

Preface

Edition 2 of *High-Yield*™ *Microbiology and Infectious Diseases* continues its tradition of acting as a review that provides for both a bug approach and a case-based infectious disease approach. All areas have been updated. In addition, the book has been slightly expanded to include the following:

1. New antimicrobials and more information on drug resistance, a hot topic on Step 1
2. Expansion of the infectious disease section to include more of the likely questions that will go with the cases

The book is still organized in such a way that you can test yourself as you go and identify areas where you need more review. In fact, if you feel like you are reasonably well-prepared, use the infectious disease tables as a self-test (use a cover sheet and note your answers on paper to make sure you are not fooling yourself, which is easily done if you just read through the tables!).

Retaining information requires active processes, including note taking, discussing, and, most important, practice in retreiving information. The latter allows faster future retrieval, which is critical for Step 1. Thus, the unique setup of the infectious disease section of this book allows you to go through most of the basic cases that will appear on the exam—something that is difficult to do with multiple-choice questions.

The material is presented using four different approaches:

1. Bug Parade Approach: Microbiology, Bacteria, Viruses, Fungi, Parasites (Chapters 1–34). These chapters cover major microbial and parasitic genera and species, presenting important organism characteristics and the diseases each organism causes. There are separate chapters covering bacterial and viral genetics. Bacterial genetic concepts are generously illustrated.
2. Organ System and Disease Approach: Infectious Diseases (Chapters 35–43). Pneumonias, rashes, and other diseases and conditions are presented in an easy-to-use clinical vignette format suitable for both fast review and self-testing. Each "high-yield" case is followed by the questions that are most likely to be asked on the exam. You won't have to flip pages for the answers—they're right below the questions! (For self-testing, use a cover sheet to cover the answers.)
3. Comparative Microbiology Approach (Chapter 44). The comparative chapter groups microorganisms according to important features—for example, "Antiphagocytic Structures" or "Toxins with ADP Ribosyl Transferase Activity." The outline format allows you to either quickly scan for review or test yourself (using a cover sheet) to ensure that when these features are used as clues on the exam, you will answer the questions with ease. This section also includes tables that highlight important toxins, culture media, and modes of transmission, which are important in differentiating infectious disease agents.
4. High-Yield Case Setup Approach (Chapter 45). This section consists of high-yield case setups, or cases composed of essential clues without the easily changeable "window dressing." The case setups are presented in random order; each case is followed by a series of likely exam questions,

with answers immediately following the questions. The format allows for both quick review and self-testing. Much faster and broader self-testing is possible with this format than with multiple-choice questions.

These four different approaches give you several options. For example, if you were taught according to an organ system approach, you might want to focus on a bug approach—or vice versa. Or you can study the entire book, which will provide the repetition you will need to retain the many facts crucial to analyzing the exam's clinical cases and answering the questions.

Other features used throughout the book to enhance your review include illustrations, tables, boldface terms, and memory tricks (identified with the Ⓜ symbol). Organisms that are particularly likely to be on the exam are often identified simply as "high-yield." In conjunction with the High-Yield™ series outline format, these features provide an efficient review.

The USMLE's (and COMLEX's) Strategy for Microbiology Testing

On the USMLE and COMLEX examinations, many questions start with a clinical scenario. For the microbiology and infectious disease questions, you will need to recognize the disease presentation and identify the causative agent based on clinical or basic science information. You may be asked to identify the causative agent, but more frequently, that question will be skipped and instead you will be asked a basic science question about the causative agent, the disease, or the therapeutic mechanisms. For instance, the exam may present a case history of a child with cough, coryza, conjunctivitis, and small gray oral lesions with a red base, which is followed by the development of a macular papular rash from the ears down that becomes confluent on the face. You might be asked to identify the family of the viral causative agent (measles is caused by Rubeola, which is a Paramyxovirus) or the replicative intermediate of the causative agent (it replicates through a positive RNA intermediate). In infections that have a dominant causative agent (e.g., a urinary tract infection), you will need to know what that causative agent is (in this case, *Escherichia coli*). In some cases, distinguishing clues will be used to point you to a specific, but less common, organism; for example, if a catalase-positive, Gram-positive coccus is the predominant organism in a urinary tract infection, then the infecting organism is *Staphylococcus saprophyticus*. Clues may include bug characteristics, virulence factors, geography, route and timing of exposure, patient's age, underlying condition, and symptomology.

There are also questions that compare different organisms or groups of organisms, toxins, or other virulence factors. As an example: A case of pneumonia (maybe without geographical or environmental clues) with a description of the causative agent in tissue, such as "broad-based budding yeast in tissue that grows as a filamentous form with nondescript small conidia at room temperature" *(Blastomyces dermatitidis)*. The next question might be about the disease's endemic regions (great riverbeds in the United States and the southeastern U.S. seaboard states) or where in the environment the exposure originated (high-organic soil with rotting wood). An association with bats would indicate *Histoplasma;* with pigeons, *Cryptococcus;* and with desert sand, *Coccidioides*. Because of the low computer screen resolution on the computerized exams, microscopic images are less likely to be used than verbal descriptions.

In the great majority of questions, high-order thinking will be required. In other words, you will have to piece together several clues. This book presents the essential database of concepts and facts that will enable you to answer the questions. The fourfold approach of *High-Yield™ Microbiology and Infectious Diseases* provides not only the repetition needed to improve retention of the data but also several options for optimizing your review. In addition, Part VI: Infectious Disease and the high-yield case setups in Chapter 45 will provide some insight and practice into problem solving.

If you have any comments about this High-Yield™ title or suggestions for the next edition, please send them to the publisher at Lippincott Williams & Wilkins, Review Books, Rose Tree

Corporate Center, Building II, Suite 5025, 1400 N. Providence Road, Media, PA 19063-2043, or via e-mail to Rounds@lww.com.

What do you need to do well on the USMLE?

High-Yield Microbiology and Infectious Disease

- Provides a way to analyze how much more you need to study microbiology and basic infectious disease.
- Summarizes the most important concepts and information you'll need in three different formats: bug parade, organ-system infectious disease, and comparative microbiology.
- Is presented in a concise manner for fast, Step 1–focused review.
- Uses mnemonics and bare-bone cases with questions to get microbiology to stick in the brain.
- Provides practice retrieving information from case-based, bug-based, or comparative questions.
- Helps increase your speed and ease of information retrieval with practice solving cases and high-yield questions.

How to Start?

1. Test yourself to assess how much more you need to study. If you feel very well-prepared, start with the cases in Part VIII. If after the first six questions, you are not doing well, then stop—you need to do more studying. (If you flew through them doing well, just test yourself with the remaining parts.) Check to see how well you know the viruses by making a list of negative RNA viral families. List which are enveloped and which are naked. Check the list against Table 23-1. Often, a case will require you to identify the causative agent by its description. Check your knowledge of virulence factors by using a cover page on material in Chapter 44. Do you know virulence factors or epidemiology? These will be used in cases on the exams.

2a. If you did fine, then continue testing yourself, using a cover sheet with Parts VI, VII, and VIII as a self-test.

2b. If you had problems, start at the beginning of the book and take notes notes about things you did not know so you can review that information sometime later. Then do Parts VI, VII, and VIII as a self-test.

I wish you good studying and good luck!

Louise Hawley, Ph.D.

Contents

Preface vii
Recurring Abbreviations xiii

I **Introduction** 1

1. From Prions to Parasites 3

II **Bacteria** 5

2. Bacterial Structure 7
3. Bacterial Metabolism and Growth 14
4. Bacterial Pathogenicity 17
5. Bacterial Genetics 27
6. Antibiotics and Drug Resistance 39
7. Identification of Major Bacterial Groups 44
8. Gram-Positive Cocci 50
9. Gram-Positive Bacilli 55
10. Mycoplasmas 62
11. Gram-Negative Cocci 64
12. Gram-Negative Aerobic Bacilli 67
13. Gram-Negative Microaerophilic Curved Bacteria 71
14. Gram-Negative Facultative Anaerobic Bacilli: Family Enterobacteriaceae 73
15. Gram-Negative Facultative Anaerobic Bacilli: Non-Enterobacteriaceae 80
16. Gram-Negative Anaerobic Bacilli and Cocci 83
17. Spirochetes (Gram-Negative Envelope) 85
18. Rickettsiaceae and Relatives 90
19. Chlamydiae 93

III **Viruses** 97

20. Viral Basics 99
21. DNA Viruses 109
22. Positive (+) RNA Viruses 116
23. Negative (–) Single-Stranded RNA Viruses 122
24. Double-Stranded RNA Viruses: Family Reoviridae 127
25. Viral Genetics 128

IV **Fungi** .133

26. Fungal Basics .135
27. Fungi That Cause Skin and Subcutaneous Infections .138
28. Systemic Fungal Pathogens .141
29. Opportunistic Fungi .144

V **Parasites** .149

30. Parasite Basics .151
31. Protozoan Parasites .155
32. Trematodes (Flukes) .163
33. Cestodes (Tapeworms) .166
34. Nematodes (Roundworms) .168

VI **Infectious Diseases** .173

35. Eye and Ear Infections .175
36. Respiratory Tract Infections .177
37. Nervous System Infections .181
38. Gastrointestinal and Hepatobiliary Disease .184
39. Cardiovascular Infections, Septicemias, and Blood Cell Changes in Infection .187
40. Bone or Joint Infections .189
41. Genitourinary Tract Infections .190
42. Cancers Associated with Infections .192
43. Skin and Subcutaneous Infections and Rashes .193

VII **Comparative Microbiology** .197

44. High-Yield Microbial Clues and Comparisons .199

VIII **High-Yield Case Setups** .205

45. High-Yield Case Setups .207

Index .219

Recurring Abbreviations

3d	3 days
3w	3 weeks
3yo	3 years-old
Ab	antibody
AFB	acid-fast bacilli
a.k.a.	also known as
Ag	antigen
BALF	bronchiole alveolar fluid
CA	causative agent
cAMP	cyclic adenosine monophosphate
CAMP	Test that identifies *Streptococcus agalactiae*'s incomplete hemolysin (C A M-P are the initials of the test developers.)
CF	cystic fibrosis
CIE	counterimmunoelectrophoresis
CMI	Cell-mediated immunity
CNS	central nervous system
COPD	chronic obstructive pulmonary disease
CSF	cerebrospinal fluid
DFA	direct fluorescent antibody
DIC	disseminated intravascular coagulation
DNA	deoxyribonucleic acid
DOC	drug of choice
DR	drug resistance or drug resistant (MDR = multiple drug resistance)
ds	double-stranded
EF-2	elongation factor 2
EHEC	enterohemorrhagic *E. coli*
EIA	electroimmunoassay
EIEC	enteroinvasive *E. coli*
ELISA	enzyme-linked immunosorbent assay
EPEC	enteropathogenic *E. coli*
ETEC	enterotoxic *E. coli*
FTA-ABS	fluorescent treponemal antibody absorption test
G^+	Gram-positive
G^-	Gram-negative
GAS	Group A Streptococcus = *Streptococcus pyogenes*
GBS	Group B Streptococcus = *Streptococcus agalactiae*
GI	gastrointestinal
GU	genitourinary
HCW	health care workers
Hib	*H. influenzae* type b
HIV	human immunodeficiency virus
HPV	human papillomavirus
HSV	herpes simplex virus

HUS	hemolytic uremic syndrome
IC	immunocompromised
IF (DFA or IFA)	immunofluorescence (Direct Fluorescent Antibody or Indirect Fluorescent Antibody stain)
IL	interleukin
IFN	interferon
IS	insertion sequence
IV	intravenous
IVDA	intravenous drug abuse or abuser
lab ID	laboratory identification
LAM	lipoarabinomannan
LPA	latex particle agglutination
LPS	lipopolysaccharide
MAC	membrane activation complex (a.k.a. serum resistance)
MALT	mucosa-associated lymphoid tissue
M cells	intestinal M cells over Peyer's patches
MDR	multiple drug resistant (or resistance)
MHC	major histocompatibility complex
MIC	minimal inhibitory concentration
mRNA	messenger RNA
MRSA	methicillin-resistant *Staphylococcus aureus*
NA	nucleic acid
NK	natural killer
OIP	obligate intracellular pathogen
OMP	outer membrane protein
PAI	pathogenicity island
PBP	penicillin-binding protein
PCR	polymerase chain reaction
PID	pelvic inflammatory disease
PMNs	polymorphonuclear neutrophil leukocytes
PrP or Prp^{sc}	prion
PPD	purified protein derivative
PrP^{C}	cellular prionlike proteins
Pt	patient
PTSAg	pyrogenic (exo)toxin superantigen
RBCs	red blood cells
RES	reticuloendothelial system
R-factor	resistance factors
RMSF	Rocky Mountain spotted fever
SPE-A	*Streptococcus pyogenes* exotoxin A
spp.	species
ss	single-stranded
STD	sexually transmitted diseases
tb	tuberculosis
TIG	Tetanus immune globulin
TNF	tumor necrosis factor
TSS	toxic shock syndrome
TSST-1	toxic shock syndrome toxin 1
UPEC	uropathogenic *E. coli*
USMLE	U.S. Medical Licensing Examination
URI	upper respiratory (tract) infection
URT	upper respiratory tract
UTI	urinary tract infection
VDRL	Venereal Disease Research Laboratory test
VZIG	varicella zoster immunoglobulin
VZV	varicella-zoster virus
WBCs	white blood cells

Part I

Introduction

Chapter 1

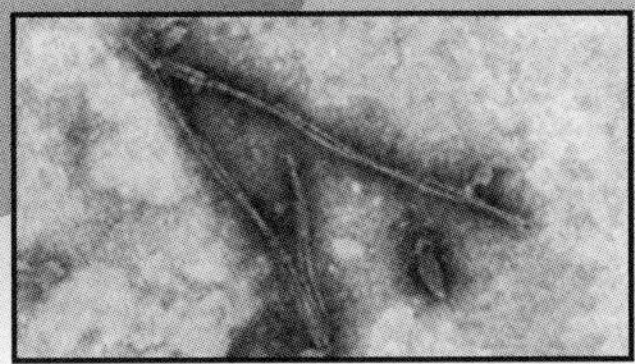

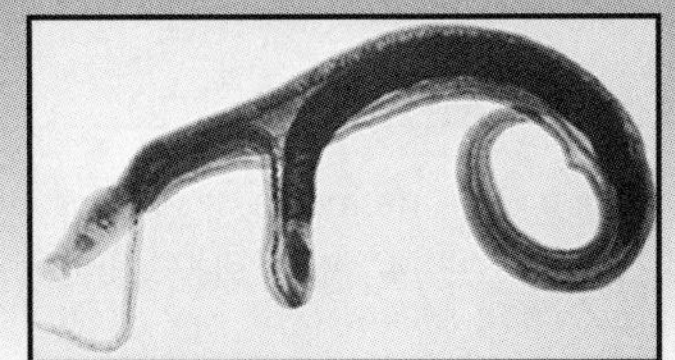

From Prions to Parasites

Microbial Groups

Each major microbial group has different characteristics, which result in distinct disease mechanisms, varying host responses, and the need for different interventions.

I Two Cell Types

Of the cellular microorganisms, there are **two different cell types: prokaryotes** and **eukaryotes.**

A. **HUMANS, ANIMALS, PARASITES*, PLANTS, AND FUNGI** are the major **eukaryotes.**

B. **BACTERIA** are the major **prokaryotes.**

C. **MAJOR DIFFERENCES** are outlined in Table 1-1.

II Microbial Groups

A. **PRIONS (PrP or PrP^{SC})** are noncellular, **infectious proteins.**
   1. **Prions are associated with subacute spongiform encephalopathies,** such as Creutzfeldt-Jakob disease, kuru, and fatal familial insomnia (or the sheep disease, scrapie.)
   2. **Prions are naked proteins that have the same amino acid sequence as certain normal human cell surface proteins but that are folded differently.** Prions are resistant to nucleases (because they have no nucleic acid), proteases (probably because of the folding), many chemicals, and normal autoclaving.

*The term *parasite* is used (1) specifically to mean the animal parasites, such as protozoans and helminths, and (2) more broadly to define an organism that derives its nutrition from another organism. In general in this text, *parasite* refers specifically to animal parasites, and *pathogen* will be substituted for the broader usage.

TABLE 1-1 PROKARYOTIC VS. EUKARYOTIC CELLULAR ORGANIZATION

Characteristic	Prokaryotic Cells (Bacteria)	Eukaryotic Cells (Fungi, Plants, Protozoans, and Animals)
Genetic organization	1 circular DNA molecule organized into multiple loops with nonhistone proteins and RNA	Linear DNA condensed with histones
	1 chromosome, multiple copies	More than 1 chromosome
	Mono- and polycistronic mRNA	Monocistronic mRNA
	Exons; no introns	Exons and introns
	No nuclear membrane	Nuclear membrane present
	Continuous DNA synthesis	G and S growth phases
Cytoplasmic structures	No mitochondria, no ER, no lysosomes	Mitochondria, ER, lysosomes
	70S ribosomes (30 and 50S subunits)	80S ribosomes (40S and 60S subunits)
Nuclear/cell division	Binary fission (asexual)	Mitosis or meiosis with cytokinesis
Cell diameter	Most <0.7μ	Most >3μ

KEY: DNA, deoxyribonucleic acid; RNA, ribonucleic acid; mRNA, messenger RNA; ER, endoplasmic reticulum.

3. **Human cells make a normal protein** (with the same amino acid sequence as the prion) on the membrane surface, which is coded for by a human cellular gene and referred to as cellular prionlike protein **(PrP^c).**
4. **Once prion proteins gain entry into the human, they modify the folding of normal PrP^c, creating additional prions,** which are released from the cellular membrane. The cell makes new PrP^c, which are then also converted. **The prions accumulate in tangles and eventually cause neurologic disease.**

B. VIRUSES are obligate intracellular organisms (i.e., they cannot be grown outside a host cell).

1. **Viruses are composed of either RNA or DNA surrounded by various proteins. Viruses may or may not have an envelope;** if so, the envelope is studded with viral glycoproteins.
2. **Viruses take over host cells** and, using the viral nucleic acid, direct the synthesis and assembly of viral components to make new viruses.

C. BACTERIA are **prokaryotic cells.**

1. Bacteria have **70S ribosomes,** complex **cell walls of peptidoglycan** (except for mycoplasmas and chlamydiae), and **no membrane sterols** (except mycoplasma/ureaplasma).
2. Bacteria divide asexually by binary fission.

D. FUNGI are **eukaryotic organisms.**

1. Fungi have **complex carbohydrate cell walls containing chitin, glucans, and mannans.**
2. Fungal **membranes have ergosterol** as the major sterol, allowing treatment with imidazoles and polyene drugs.
3. Fungi include the **yeasts, filamentous molds, dimorphic fungi,** and **mushrooms.**

E. PARASITES are eukaryotes. The parasites are the protozoans, worms, and insects that live on other organisms. **Parasites have sterols in their cell membranes but do not have cell walls.**

Part II

Bacteria

Chapter 2

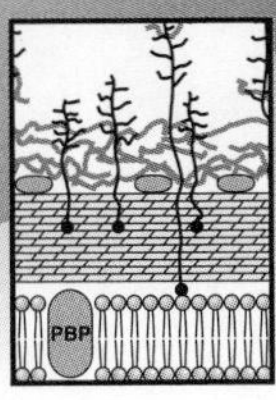

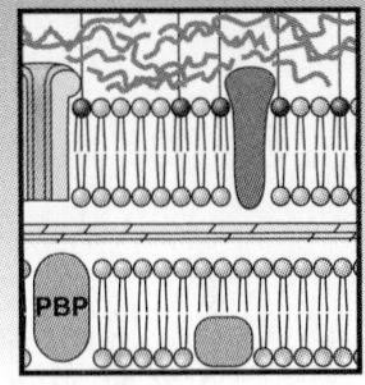

Bacterial Structure

I Cell Envelope

The cell envelope **consists of the cytoplasmic membrane, cell wall, outer membrane** (Gram-negative bacteria only), and, **for some bacteria, capsule.**

A. ROLE OF THE CELL ENVELOPE. In addition to **protecting the bacterial cell**, the cell envelope components play a major role in **adherence** to or **invasion** of human cells, **virulence**, and **stimulation of the immune response.**

B. STRUCTURES, CHEMISTRY, AND FUNCTION

1. **The components of the cell envelope, as well as the chemistry and functions of these structures, differ between Gram-positive and Gram-negative bacteria, as depicted in Figures 2-1 and 2-2.** (Note: This high-yield information is likely to be tested on the USMLE Step 1 exam.)
2. **Differences in the cross-linkage and thickness of the peptidoglycan of Gram-positive and Gram-negative bacteria lead to differential retention of the Gram dye complex** and are shown in Figure 2-3.
   a. The Gram-positive peptidoglycan "net," with its many layers and pentaglycine bridges, is more tightly cross-linked and thicker, so it retains the large dye complex inside the peptidoglycan.
   b. The Gram-negative outer membrane is damaged by the organic solvent decolorizer. The thinner, less cross-linked peptidoglycan does not retain the dye in decolorization.
   c. Table 2-1 reviews the Gram stain steps.
3. Differences between the cell envelopes of Gram-positive and Gram-negative bacteria are highlighted in Table 2-2. In general, **the Gram-negative outer cell surface is more strongly negatively charged, so these bacteria are inherently more resistant to the body's natural antibacterial peptides and to phagocytosis.** In addition, the outer membrane provides a permeability barrier restricting the entry of organic compounds through porin channels, so entry is controlled by porin channel size and porin surface charges.

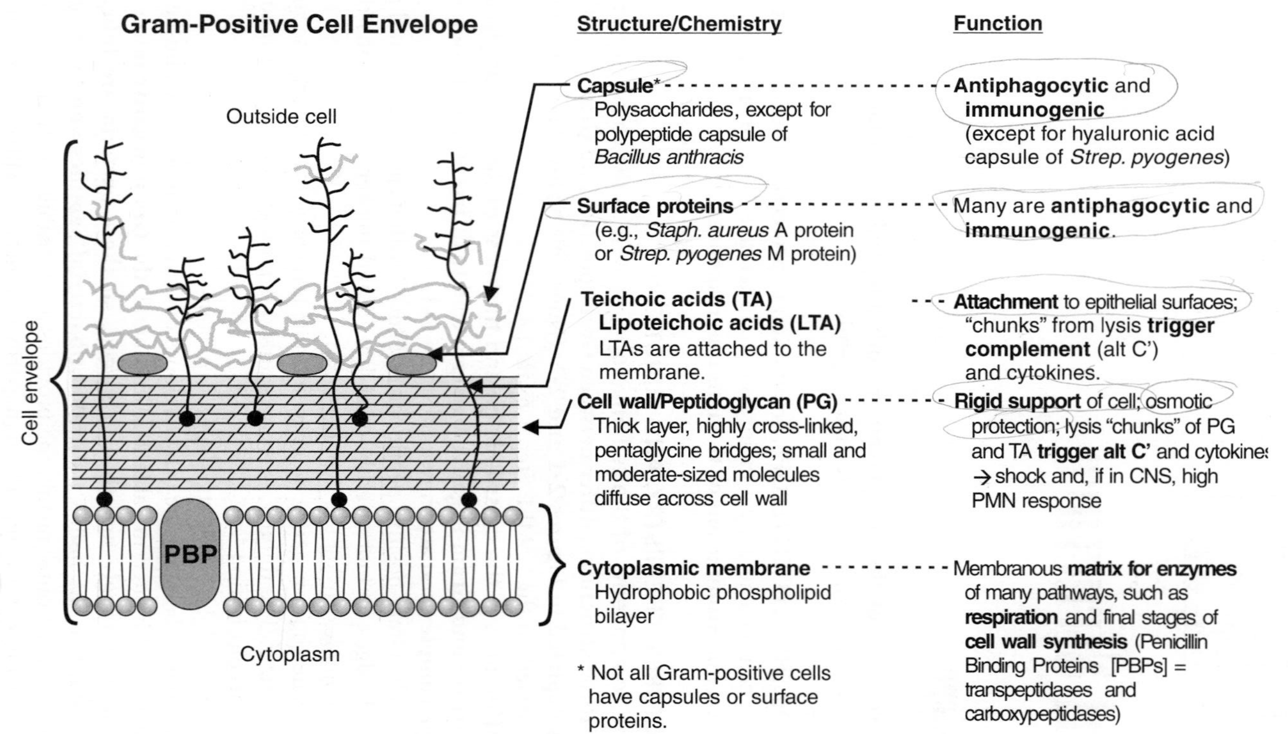

Gram-Positive Cell Envelope
Structure/Chemistry
Function
Outside cell
Cell envelope
PBP
Cytoplasm
Capsule*
Polysaccharides, except for polypeptide capsule of *Bacillus anthracis*
Antiphagocytic and immunogenic (except for hyaluronic acid capsule of *Strep. pyogenes*)
Surface proteins
(e.g., *Staph. aureus* A protein or *Strep. pyogenes* M protein)
Many are antiphagocytic and immunogenic.
Teichoic acids (TA)
Lipoteichoic acids (LTA)
LTAs are attached to the membrane.
Attachment to epithelial surfaces; "chunks" from lysis trigger complement (alt C') and cytokines.
Cell wall/Peptidoglycan (PG)
Thick layer, highly cross-linked, pentaglycine bridges; small and moderate-sized molecules diffuse across cell wall
Rigid support of cell; osmotic protection; lysis "chunks" of PG and TA trigger alt C' and cytokines → shock and, if in CNS, high PMN response
Cytoplasmic membrane
Hydrophobic phospholipid bilayer
Membranous matrix for enzymes of many pathways, such as respiration and final stages of cell wall synthesis (Penicillin Binding Proteins [PBPs] = transpeptidases and carboxypeptidases)
* Not all Gram-positive cells have capsules or surface proteins.

● **Figure 2-1** Gram-positive cell envelope showing structures and describing their chemistry and function.

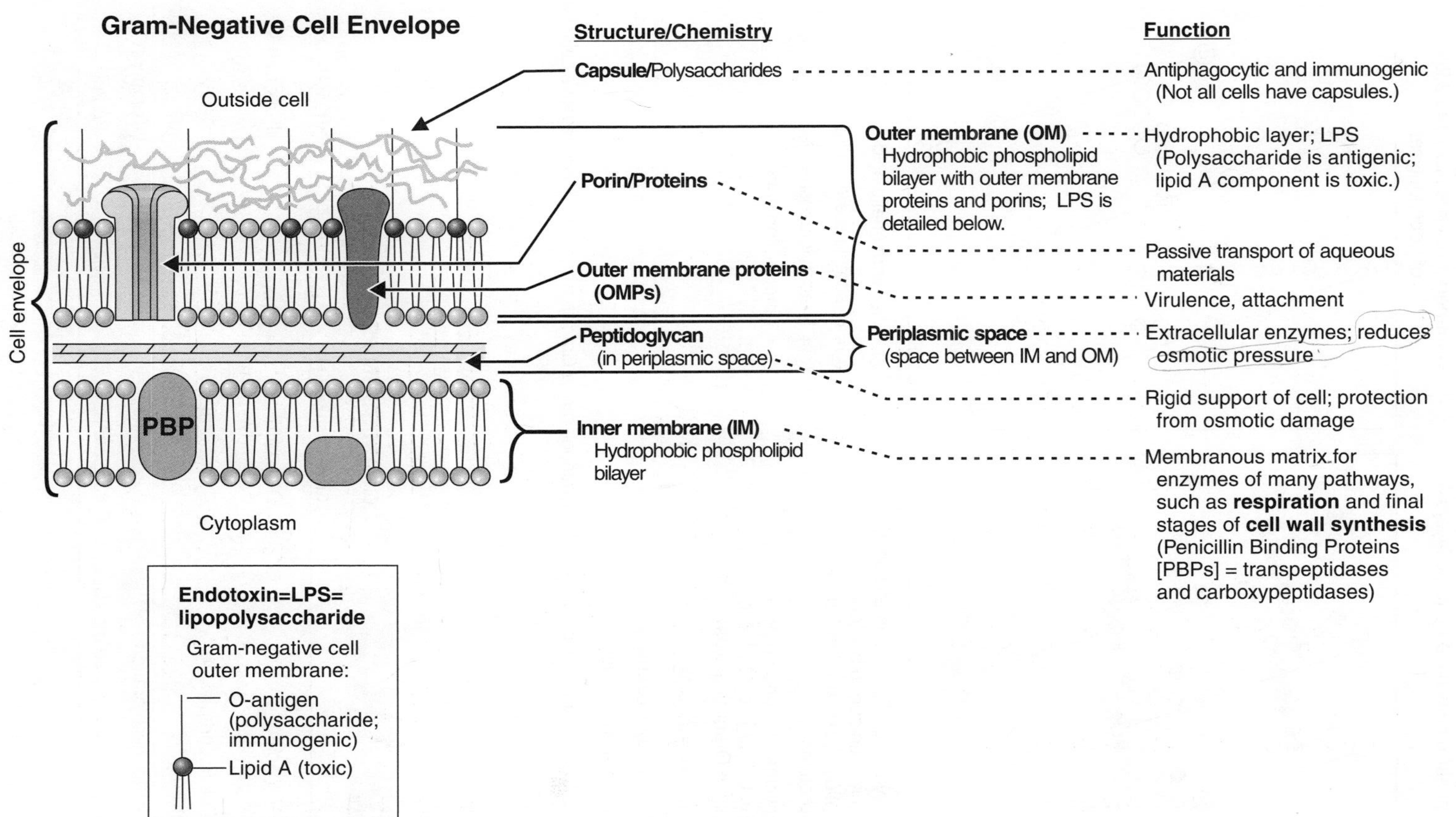
Gram-Negative Cell Envelope
Structure/Chemistry
Function
Outside cell
Cell envelope
PBP
Cytoplasm
Capsule/Polysaccharides
Antiphagocytic and immunogenic (Not all cells have capsules.)
Outer membrane (OM)
Hydrophobic phospholipid bilayer with outer membrane proteins and porins; LPS is detailed below.
Hydrophobic layer; LPS (Polysaccharide is antigenic; lipid A component is toxic.)
Porin/Proteins
Passive transport of aqueous materials
Outer membrane proteins (OMPs)
Virulence, attachment
Periplasmic space
(space between IM and OM)
Extracellular enzymes; reduces osmotic pressure
Peptidoglycan
(in periplasmic space)
Rigid support of cell; protection from osmotic damage
Inner membrane (IM)
Hydrophobic phospholipid bilayer
Membranous matrix for enzymes of many pathways, such as respiration and final stages of cell wall synthesis (Penicillin Binding Proteins [PBPs] = transpeptidases and carboxypeptidases)
Endotoxin=LPS= lipopolysaccharide
Gram-negative cell outer membrane:
O-antigen (polysaccharide; immunogenic)
Lipid A (toxic)

● **Figure 2-2** Gram-negative cell envelope showing structures and describing their chemistry and function.

Gram-Positive
Gram-positive cells have up to 60 layers.

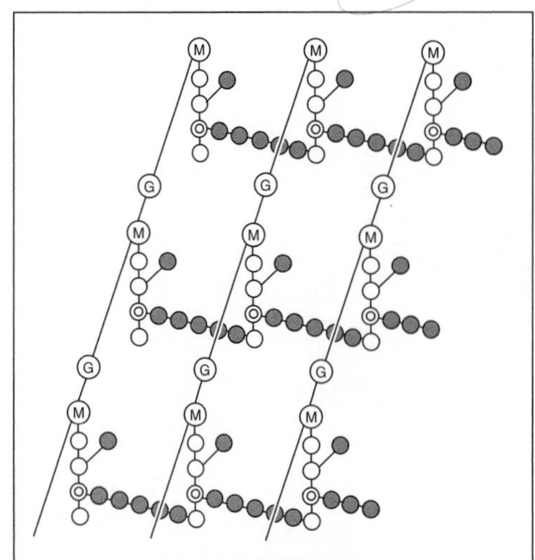

◎ = L-lysine or diaminopimelic acid, which is unique to bacterial peptidoglycan and not found in proteins

● = glycines making up penta-glycine bridge cross-linking (only in Gram-positives)

Ⓜ = N-acetylmuramic acid

Ⓖ = N-acetylglucosamine

○ = other amino acids

Gram-Negative
Gram-negative cells have 1 to 3 layers of peptidoglycan.

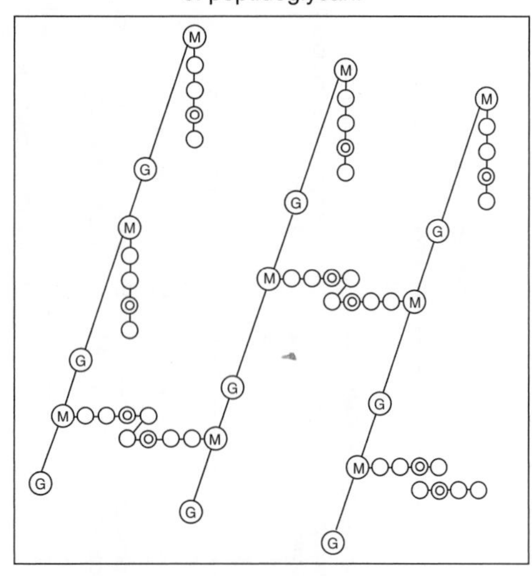

◎ = diaminopimelic acid

Ⓜ = N-acetylmuramic acid

Ⓖ = N-acetylglucosamine

○ = other amino acids

● **Figure 2-3** Comparison of cross-linkage of peptidoglycan in Gram-positive and Gram-negative bacteria.

TABLE 2-1 GRAM STAIN PROCEDURE*

Step	Gram-Positive Cell	Gram-Negative Cell
1. Gram's crystal violet (a small dye particle)	Dark purple (small dye particle)	Dark purple (small dye particle)
2. Gram's iodine (a complexing agent or mordant)	Dark purple (large dye complex)	Dark purple (large dye complex)
3. Acetone alcohol decoloration	Purple	Colorless
4. Safranin counterstain (a pale red dye)	Purple/blue	Red/pink

*Human cells Gram stain red.

TABLE 2-2 GRAM-POSITIVE VS. GRAM-NEGATIVE BACTERIA: COMPARISON OF CELL ENVELOPE FEATURES

	Gram-Positive Bacteria	Gram-Negative Bacteria
Envelope layers	Two layers: 1. Peptidoglycan (open, netlike; 1–60 layers); highly cross-linked 2. Cytoplasmic membrane (hydrophobic)	Three layers: 1. Outer membrane (OM) (hydrophobic) 2. Peptidoglycan (open, netlike; 1–3 layers) 3. Inner membrane (hydrophobic)
Unique features	Teichoic acids	OM with endotoxin Periplasmic space Porins*
Peptidoglycan (open net)	Thick (many layers), highly cross-linked	Thin layer, fewer cross-links; net has larger holes
Surface antigens	Teichoic acids Surface proteins, such as: • M protein on GAS (fimbria) • Tuberculin (*Mycobacterium tuberculosis*) • *Staphylococcus aureus* A protein Flagella Capsular material (antigenic except for hyaluronic acid (GAS))	O antigen (polysaccharide of lipopolysaccharide) OM proteins and fimbriae Flagella Capsular material (antigenic except if highly sialylated (*Neisseria meningitidis* B serotype)
Attachment to human cells	Teichoic acids Adhesins	Fimbriae Adhesins
Response to immune system	Resistant to antibody and complement-mediated killing, but susceptible to lysozyme	OM reduces susceptibility to lysozyme but is susceptible to damage by antibody and complement.
Major triggers of inflammation	Peptidoglycan-teichoic acid	Endotoxin
Effect of beta-lactam antibiotics	Antibiotic can diffuse directly through peptidoglycan and bind to PBPs, inhibiting cell wall cross-linkage. Gram-positive bacteria are usually more susceptible to beta-lactam drugs	All enter porins, cross the periplasm and peptidoglycan to reach the PBPs. Beta-lactamases (if produced), accumulate in periplasm, setting up "mine field." But beta-lactams are drug of choice for some Gram-negative bacteria, such as *Treponema pallidum* and *N. meningitidis*.

KEY: GAS, group A streptococci; PBPs, penicillin-binding proteins

*Porins are also found in the mycobacteria perforating the waxy cell layer. Mycobacteria are Gram-positive, so they do not have an outer membrane.

II Surface Protrusions

A. FLAGELLA (not found on all bacteria) **are semi-rigid, helical filaments** made up of protein. **Counterclockwise rotation produces directed motion;** clockwise rotation produces tumbling motility. A flagellum is shown in Figure 2-4.

B. PILI (fimbriae) are proteinaceous microfilaments (often with specialized tips) that extend through the cell envelope and beyond. They may be **categorized as adhesins or lectins** (binding to specific host cell receptors), **evasins** (inhibiting phagocytic uptake in the nonimmune individual), or **sex pili** (establishing the cell-to-cell contact needed for bacterial conjugation).

C. ENVELOPE SURFACE ANTIGENS include **teichoic acids,** certain **outer membrane proteins** that affect adherence or virulence (such as the ability to invade nonphagocytic host cells), and **most capsules.**

D. CAPSULES are polysaccharides that inhibit phagocytic uptake through a variety of mechanisms in nonimmune individuals.

III Interior Structures

A. LACK OF MEMBRANE-BOUND ORGANELLES. Bacteria are prokaryotic and lack membrane-bound organelles (e.g., mitochondria and lysosomes). **Respiratory enzymes and cytochromes are embedded in the bacterial cytoplasmic membrane.**

B. ENDOSPORES. Endospores occur in two Gram-positive genera of bacteria: ***Bacillus*** **(aerobic)** and ***Clostridium*** **(anaerobic).** Endospores are **resistant to killing by boiling, cold, desiccation, and antiseptics.** Figure 2-5 describes sporulation and the structure of the endospore.

C. CHROMOSOMES. Commonly, the **bacterial chromosome is a single, covalently closed circle of double-stranded DNA.** There may be multiple copies of the one chromosome.

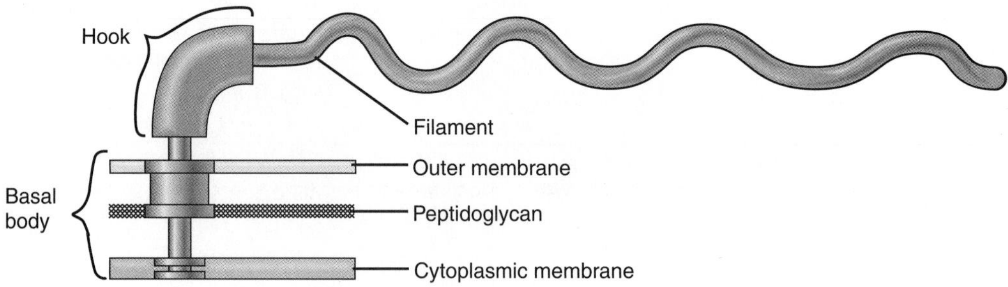

Figure 2-4 Structure of a bacterial flagellum. (The one shown is a Gram-negative bacterium.)

Sporulating Genera: *Clostridium* and *Bacillus*

Function: Survival in adverse conditions (heat, chemicals, dessication, etc.)
No increase in cell numbers

Process:

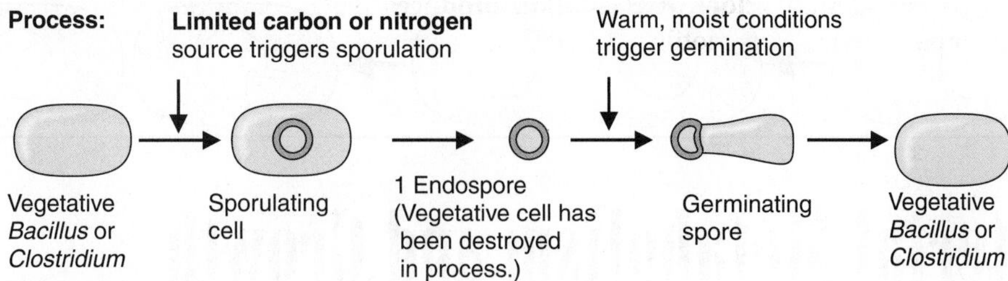

Mechanism of resistance:

Dehydration stabilizes proteins and nucleic acids.
- Calcium dipicolinate is produced by sporulating cells and plays a role in dehydration of nucleic acid.
- Calcium dipicolinate is unique to *Bacillus* and *Clostridium.*

Some new heat-resistant enzymes are produced.
Keratin coat provides protection.

Structure:

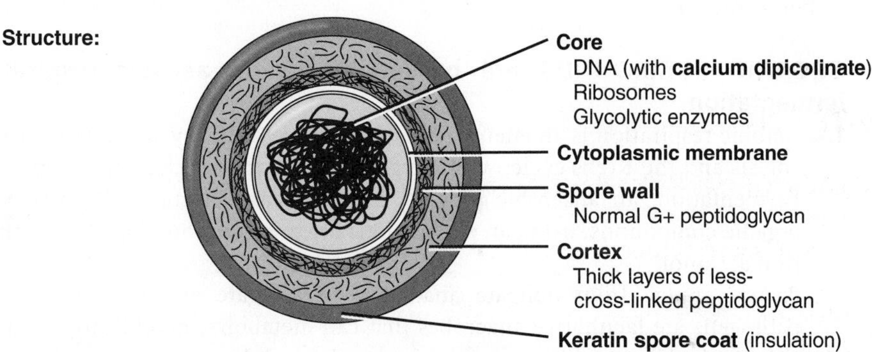

● **Figure 2-5** Sporulating process and the structure of an endospore. Sporulating genera are *Clostridium* and *Bacillus*.

Chapter 3

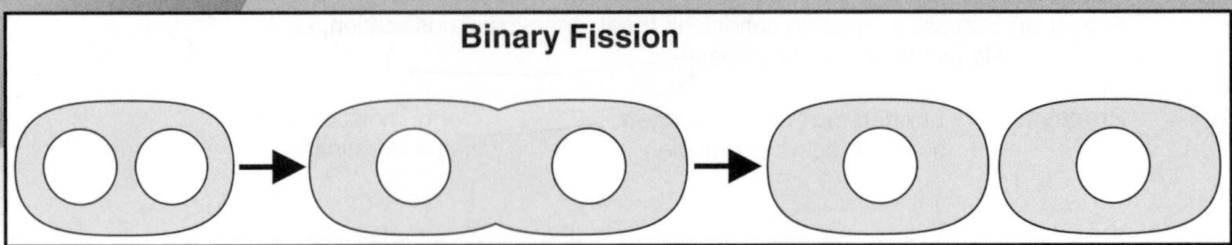

Bacterial Metabolism and Growth

I Bacterial Metabolism

A. **PATHOGENIC BACTERIA are heterotrophs** (they derive their energy from organic carbon sources).

B. **ORGANIC COMPOUNDS are broken down by aerobic respiration or fermentation.**

1. Aerobic respiration is an energy-generating process that cycles sugars through glycolysis and the Krebs cycle, with oxygen as the terminal electron acceptor.
2. Fermentation, an anaerobic process, is the release of energy by metabolism of organic compounds, using an endogenous organic compound as the terminal electron acceptor.
3. Bacteria range from obligate anaerobes to obligate aerobes. The majority of pathogens are facultative anaerobes that can metabolize aerobically depending on conditions. High-yield genera are listed in Table 3-1.

C. **DISEASE-CAUSING BACTERIA damage the living cells and then use organic compounds from the living tissues as nutrients; these organisms are called pathogens or parasites.**

1. **Extracellular pathogens grow outside of cells.** Most can be routinely cultured in the laboratory on inert media. (Exception: Because *Treponema pallidum* cannot be cultured, it is considered an obligate pathogen.)
2. **Facultative intracellular pathogens are usually found inside cells** in the body, although they can be grown in the laboratory.
3. **Obligate intracellular pathogens grow only inside cells** and cannot be cultured in vitro (e.g., *Mycobacterium leprae,* chlamydiae, and rickettsias).

D. **SOME NONPATHOGENIC BACTERIA** (called **commensals**) **can live in or on people without causing significant disease.** Long-term commensals of the human body are often referred to as **normal flora.**

TABLE 3-1 AEROBES AND ANAEROBES

Type of Organism	Definition	High-Yield Examples
Anaerobes	Utilize fermentative pathways. Most anaerobes lack superoxide dismutase and catalase. • **Obligate anaerobes** only ferment and are killed by O_2. • **Aerotolerant anaerobes** only ferment but can tolerate oxygen. (The streptococci are so aerotolerant that they are sometimes described as "indifferent.")	*Actinomyces* *Bacteroides* *Clostridium**
Facultative anaerobes	Have both aerobic respiration and fermentation (aerobes with the "faculty" to switch to anaerobic pathways in the absence of oxygen).	Most human pathogens (e.g., enterobacteria)
Microaerophiles	Require oxygen at low levels (~5% O_2).	*Campylobacter* *Helicobacter*
Obligate aerobes	Have only aerobic respiration and no anaerobic pathways.	*Pseudomonas* *Mycobacterium tuberculosis* *Bacillus* (some species)

Ⓜ The **ABCs of anaerobiosis** are ***A**ctinomyces*, ***B**acteroides*, and ***C**lostridium*. You do not need to know which species are obligate and which are aerotolerant; just learn them as anaerobes. These three genera are critical! If you are going for a top score, learn the mnemonic—**the ABCs of anaerobiosis for FP³s**—and add ***F**usobacterium*, ***P**revotella*, ***P**orphyromonas*, and ***P**eptostreptococcus*.

II Bacterial Growth

Bacterial growth is a coordinated process of increase in individual cell mass and size and duplication of the chromosome, followed by cell division.

A. BACTERIAL REPRODUCTION involves duplication of DNA without the addition of outside DNA; thus, **bacterial cell division is asexual and gives rise to genetically identical cells.** This process is called **binary fission**, because one cell always gives rise to two cells (Figure 3-1A).

B. BACTERIAL GROWTH is measured by two basic methods:

1. **Viable counts (colony counts) enumerate only those bacteria that can give rise to a colony.** Viable methods give living cell number, not size.
2. **Nonviable methods** (e.g., optical density of the culture, dry cell mass, and quantitative measurement of an individual component) **do not distinguish dead from living cells.**

C. A GROWTH CURVE measures viable bacteria in a broth medium over time. Figure 3-1A shows the four stages of a growth curve and the events of each stage. Figure 3-1 also presents a typical USMLE Step 1 problem to the right of the growth curve. To test yourself, cover Figure 3-1B (below the heavy line) before reading the problem. **Remember, there is only one lag phase, and when the cells begin to divide, one cell always divides into two in each generation.** The answer and explanation for the problem are shown in Figure 3-1B.

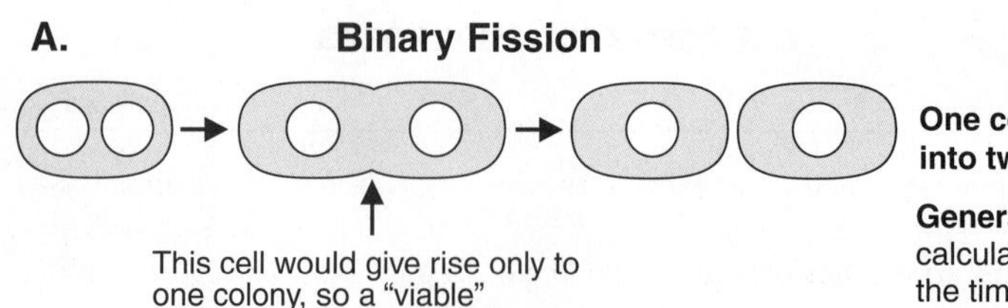

A.
Binary Fission
This cell would give rise only to one colony, so a "viable" count = 1. With optical density or cell mass, it would look like 2.

One cell always divides into two cells!

Generation time (always calculated in log phase) is the time for one cell to divide into two.

1→2→4→8→16→32→64→128→etc.

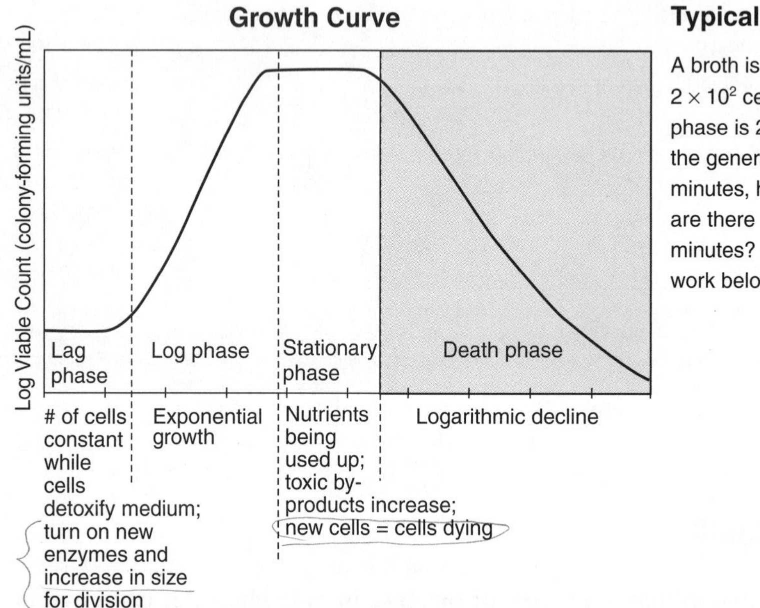

Typical Problem:

A broth is inoculated to 2×10^2 cells/mL. If the lag phase is 20 minutes and the generation time is 10 minutes, how many cells are there at the end of 60 minutes? (Answer and work below.)

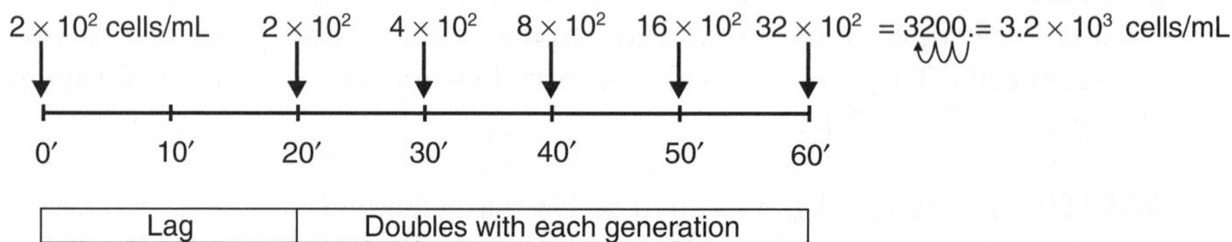

● **Figure 3-1** Bacterial growth and division. (A) Binary fission (which is asexual), growth curve, and typical problem. (B) Explanation of growth curve problem.

Chapter 4

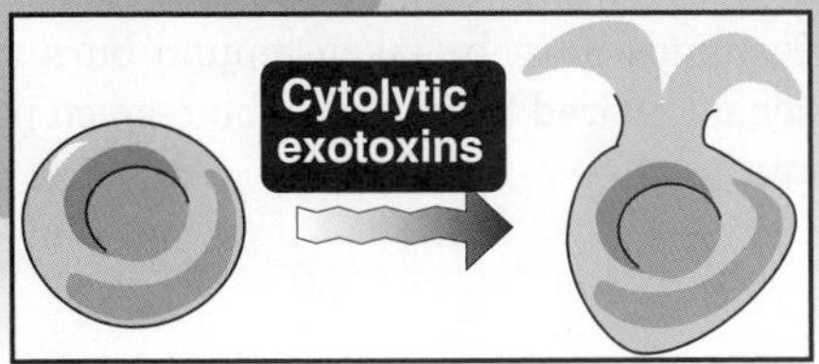

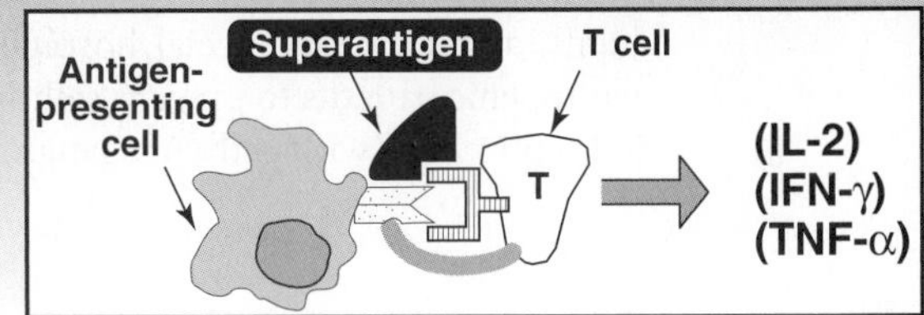

Bacterial Pathogenicity

Microbial ability to cause disease is complicated and varied. The resulting disease depends on the natural history of the organism, the amount and type of exposure, and the microbe's pathogenic mechanisms, as well as the health of the exposed person.

I Survival Outside the Human Host

Many human bacterial pathogens do not survive long in the environment. In general, these more delicate pathogens (and some not-so-delicate ones) are fairly directly transmitted by respiratory droplet spread, direct contact, or living vectors. One or more of the following factors may promote survival of the hardier pathogens:

A. RESISTANCE TO DRYING. Spore forms of ***Clostridium*** and ***Bacillus,*** the fairly inert **environmental forms of *Coxiella* and *Chlamydia*,** and **mycobacteria,** with their **waxy cell walls,** all survive drying and may be airborne.

B. COLD GROWTH. Most fecal pathogens die (somewhat slowly) in the cold. However, ***Listeria monocytogenes*** and ***Yersinia enterocolitica*** (both found in animal gastrointestinal tracts) not only survive outside the intestinal tract in the environment but actually grow at refrigerator temperature.

C. SHORT-TERM SURVIVAL SKILLS. *Staphylococcus aureus* and ***Streptococcus pyogenes*** are two notable pathogens that can survive for short periods of time on fomites. *S. aureus* is spread in hospitals, not only by direct contact but also on laundered sheets (unless bleached) and other fomites. Some strains of *Escherichia coli* (part of our normal flora) with decent short-term survival skills have unfortunately picked up some virulence genes and now cause water- and food-borne disease. In addition, most organisms survive a while on hard surfaces, such as faucets, which can be contaminated by one person who improperly washes his or her hands.

D. GROWTH AND SURVIVAL IN WATER

1. ***Legionella,*** which grows in amebae in streams, contaminates air-conditioning cooling tanks and is then spread by air-conditioned air. It has also contaminated hot water tanks and dental clinic water lines.

2. The opportunistic ***Pseudomonas*** grows well in tap and distilled water and may be found on or in raw vegetables, cut flowers, soil, hot tubs, whirlpools, faucet aerators, drains, and so forth. Special hospital precautions must be taken around burn and neutropenic patients to prevent cellulitis (characterized by odiferous blue-green pus) and septicemia (with ecthyma gangrenosum.)

II Colonization of the Human Body

A. NORMAL FLORA (a.k.a. residential microbiota). Normal flora colonizes the human body's internal and external surfaces. These organisms have low virulence and are usually kept in balance by our surface defense mechanisms.

1. Normal flora **reduces the risk that pathogens will colonize** by modifying pH, blocking binding sites, competing for nutrients, and producing antibacterial compounds called *bacteriocins* or *toxic metabolites*.
2. **Infection may result if normal flora is carried into normally sterile tissues** (e.g., by trauma or through bowel defects in patients on cytotoxic chemotherapy), which is particularly serious in immunocompromised patients.

B. COLONIZATION PATTERNS BY PATHOGENS

1. **Colonization, with local or systemic spread to cause disease.** *Streptococcus pneumoniae* colonizes the oropharyngeal mucosa. It is more likely to reach someone's lungs if the cough reflex is suppressed (e.g., in alcoholics) or if the mucociliary elevator is damaged (e.g., by smoking or following influenza or measles). Other pathogens that regularly colonize mucosal surfaces and that may cause more invasive disease are *Neisseria gonorrhoeae, N. meningitidis,* and *Haemophilus influenzae* type b.
2. **Colonization, with little invasiveness but toxin production/circulation.** Some bacteria (e.g., *Corynebacterium diphtheriae*) colonize mucosal surfaces and produce exotoxins, which circulate and cause disease.
3. **Colonization by a pathogen without disease (carrier state).** A carrier is a colonized or infected individual (usually immune) who harbors a pathogen but is not ill and who serves as a source to spread the organism to others. Carriers usually have partial immunity.

C. MAJOR MECHANISMS OF COLONIZATION

1. **Specific attachment to cellular surfaces** through receptor-ligand interaction with bacterial surface molecules.
   a. Gram-positive bacteria attach to cell surfaces primarily through the **teichoic acids.**
   b. Gram-negative bacteria attach primarily through proteins on the tips of **pili** and, in some cases, through certain outer membrane proteins.
2. **Adherence by adhesins.** Pathogens achieve tighter binding to host cells through bacterial surface proteins called *adhesins,* which are found on the outer surfaces. (Many of these adhesins are lectins, or proteins binding to carbohydrates on the host cell surfaces.) Adhesins are important in countering the flushing activity of mucosal surfaces. Strains of *E. coli* that cause pyelonephritis bind tightly to the uroepithelium with P pili and remain attached, despite the urine's flushing action.
3. **Avoiding mucin entrapment.** Mucin flows from the goblet cells in streams so that mucus is not uniform on mucosal surfaces. Secretory immunoglobulin

A (sIgA), produced by the host, binds through the antigen-binding site to bacteria. The Fc fragment then adheres to mucin, thus entrapping the bacteria in mucin. Pathogens colonizing mucosa produce an IgA protease, which cleaves sIgA, thus avoiding ensnarement and the subsequent flow out of the body; this enhances the pathogens' opportunity to adhere to the mucosa.

4. **Biofilm formation.** Biofilms are aggregates of bacteria and their slimes on surfaces. The bacteria are not evenly distributed in biofilms; microscopically, the biofilms look a bit like bacteria growing in columns of stalactites and stalagmites in a flooded cave, with some planktonic organisms floating in between. At least one of the bacteria involved produces an extracellular carbohydrate matrix (high yield: *Staphylococcus epidermidis* on catheter tips and other invasive devices; *Streptococcus mutans* in dental plaque). Biofilms are more common in the body than was earlier thought. Some antibiotic treatment failures occur because the organisms in the interior of those "columns" may not be actively growing, so they are protected from antibiotics that kill only growing cells (e.g., beta-lactam drugs). An additional medical problem is that many of our diagnostic procedures, such as blood cultures, only detect planktonic bacteria. Special techniques have been designed to culture catheter tips.
5. **Antigenic variation.** Through a cassette mechanism similar to human antibody formation, bacteria with the ability to constantly change surface antigens reduce the effectiveness of the specific immune system. ***N. gonorrhoeae*** is one of these organisms.

III Entry into Human Cells and Tissues

A. INTACT SKIN is a major barrier to the entry of bacteria. Traumatic implantation (e.g., injury, arthropod or animal bites, and surgery) may allow direct entry of normal flora (skin, mucosal, or fecal), environmental organisms, or pathogens.

B. MUCOSA SURFACES may be invaded, both with and without trauma.

1. Although healthy people may be colonized with *N. meningitidis,* in outbreaks in Africa, central nervous system (CNS) disease occurs in dry season. When the rainy season comes, there is a dramatic decrease in new cases, suggesting that the poorer hydration of the oral pharyngeal mucous membranes may play a role.
2. **Invasins** are bacterial proteins that **trigger the host** cytoskeleton to rearrange, causing phagocytic cells, as well as cells that are not normally phagocytic, **to ingest the bacteria.** The following section describes ways some pathogens then sneak around so as not to be killed.
3. **M cells,** found over Peyer patches in the gastrointestinal tract, serve as the entry point for such organisms as *Salmonella, Shigella,* and *Yersinia pseudotuberculosis.*

IV Evasion of the Immediate Host Defense System

A. AVOIDING PHAGOCYTIC UPTAKE. Ability to evade phagocytic uptake in a nonimmune host is mediated by surface structures. All bacterial **capsules** or slime layers, as well as **some pili** (notably *N. gonorrhoeae*), are antiphagocytic until opsonized.

B. EVADING PHAGOCYTIC KILLING. Certain organisms may be phagocytized but not killed.

1. *L. monocytogenes* **exits the phagosome to the cytoplasm quickly enough to avoid damage.** The human cytoplasm is a protected place where the bacteria can replicate hidden from antibodies, complement, and antibiotics that do not penetrate human cells, including many of the beta-lactams.
2. The sulfolipid of *Mycobacterium tuberculosis* (sulfatide) **inhibits the phagolysosomal fusion,** which keeps *M. tuberculosis* from being killed. Even if the fusion takes place, the waxy nature of the cell's surface reduces bacterial death rates.

C. INHIBITING CILIASTASIS OR KILLING CILIATED RESPIRATORY TRACT CELLS. *Mycoplasma pneumoniae* binds to cilia through P1, an adhesin, causing ciliastasis and, ultimately, desquamation of ciliated respiratory tract cells. This, in turn, allows growth of the organism into the lungs, which leads to pneumonia.

D. RESISTING COMPLEMENT-MEDIATED KILLING (a.k.a. membrane activation complex, or MAC). This occurs with some Gram-negative bacteria and is sometimes referred to as *serum resistance*. MAC resistance may be a result of changes in O antigens of the outer membrane or of sialylation of lipopolysaccharide (LPS).

Production of Toxins

A major mechanism by which extracellular bacteria produce disease is the production of structural toxins or protein exotoxins and/or extracellular enzymes. Details and symptoms (which you will need for case history questions) are shown in Table 4-1.

A. STRUCTURAL TOXINS. Endotoxin, produced by Gram-negative bacteria, is the best-studied structural toxin, because Gram-negative infections continued to be a problem for several decades following the introduction of penicillin. Due to acquired antibiotic resistance, about half of all cases of septic shock are now triggered by Gram-positive bacteria. When bacteria are limited to a tissue, inflammation results from the presence of these bacterial components; but when high numbers of bacteria invade the bloodstream, septic shock, often difficult to control, ensues.

1. **Endotoxin.** Endotoxin, or LPS, is released from the Gram-negative outer membrane of dying cells. Lipid A activates macrophages to produce interleukin-1 (IL-1), IL-6, IL-8, tumor necrosis factor (TNF) alpha, and platelet-activating factor. These cytokines ultimately stimulate production of prostaglandins and leukotrienes and activate both the complement and coagulation pathways. This results in an increased number of polymorphonuclear neutrophil leukocytes (PMNs) in the blood, fever, and increased respiratory and heart rates.
2. **Peptidoglycan and teichoic acids (including the lipoteichoic acids).** In infections with Gram-positive pathogens, the large fragments of peptidoglycan and teichoic acids from the cell envelope trigger inflammation and shock by activating the same chain of events as endotoxin.

B. EXOTOXINS. These proteins are usually secreted by living pathogenic bacteria, but a few build up in the cell and are released on lysis. Some directly inject a toxin into a host cell, thus avoiding any neutralizing antibodies present in the extracellular environment. An example of this **type III secretion** system is used by *Legionella* to inject Vac A toxin directly into the human cell.

TABLE 4-1 MAJOR BACTERIAL TOXINS: MECHANISMS OF ACTION AND EFFECTS

	Organism(s)	Toxin(s)	Mechanism of Action	Disease Effect
STRUCTURAL TOXINS—RELEASED ON DEATH OF CELL				
Mediators of inflammation in tissues and of septic shock when in bloodstream	Gram-negative bacteria	LPS = endotoxins; Toxic part is lipid A.	Stimulates host cells to release cytokines IL-1, TNF alpha, IL-6, IL-8, and platelet-activating factors, leading to activation of complement, coagulation, and endothelial damage.	Fever, increased respiration and pulse rates, petechial rash leading to ecchymosis, hypotension, thrombocytopenia, hemorrhage, and vascular collapse (acute respiratory distress syndrome, disseminated intravascular coagulation, and multiple organ system failure)
	Gram-positive bacteria	Peptidoglycan-teichoic acids. Disintegrating cell wall components trigger release of same cytokines as LPS.	Similar to LPS	Similar to LPS
EXOTOXINS—COMMONLY SECRETED BY LIVING CELLS				
Membrane-disrupting toxins	*Staphylococcus aureus*	Alpha-toxin	Pore-forming cytolysin	Tissue damage, hemolysis
	Streptococcus pyogenes	Streptolysin O	Pore-forming cytolysin	Tissue damage, hemolysis
	Listeria monocytogenes	Listeriolysin O	Pore-former in phagosome rapid egress from phagosome	Survives in human phagocytes, allowing it to spread
	Clostridium perfringens	Alpha-toxin	Lecithinase hydrolyzes eukaryotic phospholipid.	Tissue destruction in tissue infections

(Continued)

TABLE 4-1 MAJOR BACTERIAL TOXINS: MECHANISMS OF ACTION AND EFFECTS (*Continued*)

	Organism(s)	Toxin(s)	Mechanism of Action	Disease Effect
EXOTOXINS—COMMONLY SECRETED BY LIVING CELLS				
Superantigens	*S. aureus*	Toxic shock syndrome toxin-1 (TSST-1)	TSST-1 enters bloodstream; binds T cell receptors, nonspecifically activating large numbers of T cells and triggering massive IL-1 and TNF alpha production.	Rash, desquamation of palms and soles, hypotension, capillary leakage, multiorgan failure
	Streptococcus pyogenes	*Streptococcus pyogenes* exotoxin-A (SPE-A) (also called erythrogenic toxin or pyrogenic toxin)	Same as for TSST-1; produced only by the most virulent strains	Same as TSST-1, with added cardiotoxicity
	S. aureus	Enterotoxin B	Binds T cell receptors	Vomiting and diarrhea
A-B toxins that inhibit protein synthesis	*Corynebacterium diphtheriae*	Diphtheria toxin	Binds to and does greatest damage to epithelium, heart, and nerves. The A component ADP-ribosylates eukaryotic EF-2, inhibits protein synthesis, and kills cells.	Sore throat, slight increase in temperature, pseudomembrane in throat, neurologic and cardiac symptoms, death by asphyxiation or heart failure
	Pseudomonas aeruginosa	Exotoxin A	ADP-ribosylates EF-2, shutting down protein synthesis primarily in liver cells.	Jaundice and increased risk of death in septicemia with exotoxin A–producing strains of *P. aeruginosa*

	Shigella dysenteriae type 1	Shiga toxin (ST)	The A component cleaves 60S ribosomes, stopping protein synthesis and killing cells.	Most strains of shigella cause dysentery by invasion. But *Sh. dysenteriae* type 1 strains also make shiga toxin and do more damage to the colonic mucosa; may also cause hemolytic uremic syndrome (HUS).
	Enterohemorrhagic *Escherichia coli*	Verotoxin (a shigalike toxin)	Same as shiga toxin (above)	Hemorrhagic colitis and HUS
A–B toxins that increase cAMP	*Vibrio cholerae*	Cholera toxin	Internalized A component ADP-ribosylates G_S, which activates an adenyl cyclase, thus producing high level of cAMP.	Major fluid/electrolytes loss; clear stools with flecks of mucus (rice water stools); major dehydration if not replaced; hypovolemic shock
	Enterotoxic *E. coli* (ETEC)	*E. coli* labile toxin (LT)	As above, but not as severe	Watery diarrhea that may look like rice water
	Bacillus anthracis	Anthrax toxin	Three-component toxin with protective antigen (PA) serving as the B component for either the edema factor (EF) or the lethal factor (LF). The EF is an adenyl cyclase that activates inside cells; the LF kills.	In skin infection, malignant pustules develop with dark red vesicles (appearing tumorlike). They become black and necrotic from the center out. In respiratory disease, there is septic shock.

(*Continued*)

TABLE 4-1 MAJOR BACTERIAL TOXINS: MECHANISMS OF ACTION AND EFFECTS (*Continued*)

	Organism(s)	Toxin(s)	Mechanism of Action	Disease Effect
	Bordetella pertussis	Pertussis toxin	Pertussis toxin plays a role in the attachment of *B. pertussis* to respiratory mucosa. It is also internalized and inhibits G_i (the negative regulator of adenyl cyclase) through ADP-ribosylation, resulting in an increase in cAMP.	Increase in respiratory secretions and mucus, decreased phagocytic function in upper respiratory tract, and encephalopathy. The toxin may not be directly responsible for the paroxysmal cough associated with pertussis.
Neurotoxins	*Clostridium tetani*	Tetanus toxin	Acts on CNS, inhibiting release of inhibitory transmitters like GABA.	Rigid spasm
	Clostridium botulinum	Botulinum toxin	Acts on peripheral synapses, blocking release of neurotransmitters.	Flaccid paralysis

KEY: LPS, lipopolysaccharide; IL, interleukin; TNF, tumor necrosis factor; ADP, adenosine diphosphate; EF-2, elongation factor 2; cAMP, cyclic adenosine monophosphate; CNS, central nervous system; GABA, gamma-aminobutyric acid.

1. **Superantigen exotoxins.** Without entering macrophages and being processed as antigen, superantigens bind to MHC class II receptors on antigen-presenting cells and cross-link to T cell receptors, nonspecifically activating one in five T cells, as opposed to the normal 1 in 10,000. This causes a tremendous excess production of IL-1 and TNF-alpha, which leads to shock. The best known of the superantigen toxins are ***Staphylococcus aureus* toxic shock syndrome toxin 1 (TSST-1)** and ***Streptococcus pyogenes* exotoxin A (SPE-A),** both of which are produced only by more virulent strains under certain conditions.
2. **Membrane-disrupting bacterial exotoxins (mainly cytolysins)**
   a. **Pore-forming exotoxins** are protein toxins that are inserted into cholesterol-containing host membranes to create pores that kill the human cells.
      i. ***S. aureus* alpha-toxin** and ***Strep. pyogenes* streptolysin** are both pore-forming cytolysins that destroy a variety of cells, including red blood cells. This releases nutrients for the bacteria.
      ii. **Listeriolysin O** is a pore-forming toxin secreted by *L. monocytogenes* and activated in the phagosomal environment. It allows *L. monocytogenes* to quickly exit the phagosome into the cytoplasm, so it is undamaged by lysosomal contents.

   b. **Membrane-hydrolyzing enzyme. *Clostridium perfringens* alpha-toxin,** another cytolysin, is **a lecithinase** that hydrolyzes the eukaryotic membrane phospholipid, destabilizing cytoplasmic membranes. (Alpha-toxin and a second hemolysin called delta-toxin produce the characteristic double zone of hemolysis of *C. perfringens* on blood agar plates.)
3. **Classic A-B component bacterial exotoxins**
   a. Ⓜ One or more **B** components **b**ind to specific cell surface receptors and initiate the internalization of the A component. Thus, the B component determines what cell type each toxin damages.
   b. Ⓜ The **A** portion of the toxin is the **a**ctive toxic portion, which is internalized and activated and then inhibits a specific, critical intracellular function, thus damaging the cell.
   c. There are **three major classes of A-B toxins:**
      i. **Inhibitors of protein synthesis**
         (a) **Diphtheria** and ***Pseudomonas* exotoxin** A toxins both inhibit elongation factor 2 [EF-2].
         (b) **Shiga toxin** (including that of enterohemorrhagic *E. coli*) cleave the eukaryotic 60S ribosomal subunits rRNA.
      ii. **Toxins that increase cyclic adenosine monophosphate (cAMP)**
         (a) ***E. coli* labile toxin (LT)**
         (b) **Cholera toxin**
         (c) **Anthrax toxin**
         (d) **Pertussis toxin**
      iii. **Neurotoxins** (both endopeptidases)
         (a) **Tetanospasmin, producing rigid spasm**
         (b) **Botulinum toxin, producing flaccid paralysis**

See Table 4-1 for summary.

VI Intracellular Growth

Intracellular growth gives bacteria much protection from the immune system and partial resistance to antibiotics that do not penetrate human cells well. Bacteria may damage the host cells by limiting essential components or by interfering with the host cell's respiration.

A. BACTERIA THAT OVERRIDE KILLING BY MACROPHAGES and then grow inside macrophages include **rickettsias, mycobacteria, *Brucella* spp., and *L. monocytogenes*.**

B. SHIGELLAE INVADE THE PEYER PATCHES through the M cells, survive, and exit on the underside. Some shigellae are picked up there and killed by macrophages, but others attach to cellular integrins (only on the basal surface) and invade the columnar mucosal cells. They can then spread laterally by polymerization of actin filaments, which propels them into adjoining cells and creates shallow, lateral ulcers.

C. OBLIGATE INTRACELLULAR PATHOGENS (OIPs) include ***Mycobacterium leprae* and all members of the genera *Chlamydia*, *Rickettsia*, and *Ehrlichia*.**

D. *TREPONEMA PALLIDUM* cannot reliably be grown outside humans; it penetrates some human cells, so it is an obligate pathogen but not an OIP.

Chapter 5

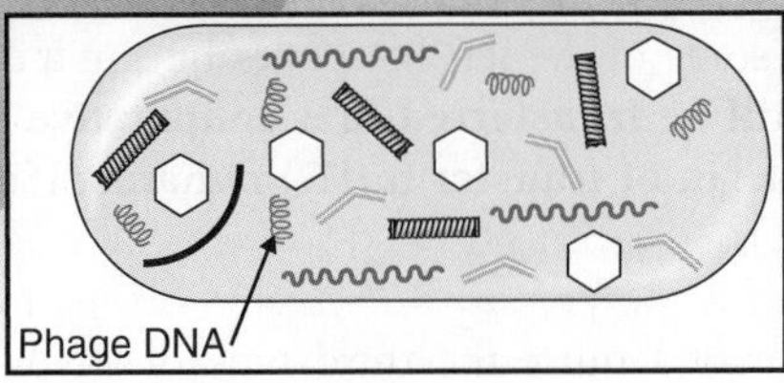

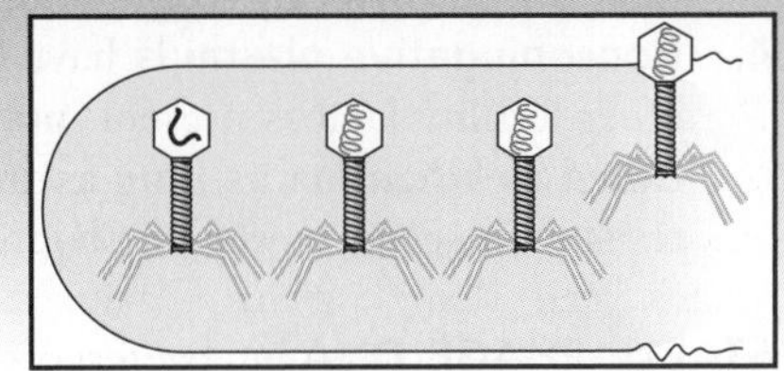

Bacterial Genetics

 DNA Molecules in Bacterial Cells

Genes expressed by bacterial cells may be located on one of three types of DNA molecules: **the bacterial chromosome**, **plasmid DNA**, or **phage DNA.** Special gene segments called **transposable elements** (including transposons and insertion sequences) have the ability to move themselves (or a copy) around on DNA or from one DNA molecule to another (e.g., from plasmid to bacterial chromosome or vice versa). These elements have been important in creating multiple drug resistance plasmids and also play a role in inversions of DNA segments, which may lead to antigenic or phase variation by moderating gene expression (e.g., *Neisseria gonorrhoeae* surface antigens). **Pathogenicity islands (PAIs)** are sections of DNA that have sufficiently different GC ratios that they probably originated in another organism. *Helicobacter*'s PAI carries a type III secretion system and many of the virulence genes.

A. BACTERIAL CHROMOSOME (always present). This is most commonly a single, large, covalently closed molecule (~3,000 times the diameter of the cell) of typical double-stranded DNA. It is apparently organized into loops and condensed to fit into about 50% of the cell volume. It has **no major histones**, and there is **no nuclear membrane** separating it from the cytoplasm. **It contains all essential bacterial genes.** There may be more than one copy in a cell. Due to the rapid replication of most bacteria and their haploid nature, mutations are rapidly expressed.

B. PLASMID DNA. Plasmids are **extrachromosomal pieces of DNA** commonly found in bacteria. They are "circular," like bacterial chromosomes, but much smaller. There may be more than one kind in the cell, as well as more than one copy for any of them. They may also be lost from the cell.

1. **Plasmid functions.** Under most growth conditions, **plasmid genes are nonessential, but they may be important to pathogenicity** (e.g., through production of **some bacterial toxins**) or to survival under special conditions (e.g., some **antibiotic resistance genes** in the presence of that antibiotic).
2. **Episomal versus nonintegrative plasmids**
   a. **Episomes** are a subclass of plasmids that can **insert themselves into the bacterial chromosome.** (This requires site-specific recombination, to be discussed later.) This ability ensures replication of the plasmid with the bacterial DNA and equal distribution to daughter cells.
   b. **Nonintegrative plasmids remain free** and may be lost in cell division.

3. **Conjugal plasmids** (e.g., fertility factors) have a series of complex genes to **direct their own transfer by conjugation.** (They also may transfer attached bacterial DNA.)
4. **Nonconjugative plasmids** have lost some of the fertility genes but, in a cell with other fertility factors present, **may still be transferred in a conjugative process called *mobilization* as long as the origin of transfer (*oriT*) remains.** (See drug-resistant *N. gonorrhoeae* in Chapter 6.)

C. **STABLE PHAGE DNA.** A bacterial cell may acquire the third type of DNA—phage (a.k.a. bacterial virus) DNA—through infection with a temperate phage whose viral production is repressed (**lysogeny,** detailed in Section IV.C) or through infection with a defective phage that cannot make phage but may replicate the DNA and express gene products. The latter is the situation with botulinum toxin.

II Mutation

As with other biological systems, mutations occur. Due to the rapid replication of most bacteria and their haploid nature, mutations are rapidly expressed. These mutations include insertion, deletion, substitution, and inversions (often a consequence of insertion sequences.)

III Homologous Recombination

Homologous recombination is the stabilization of newly introduced bacterial genes coming in on linear pieces of DNA. When new bacterial genes enter a bacterial cell by any gene-transfer process (see Section IV.A), they are imported as linear pieces of DNA (exogenotes). To avoid rapid degradation by the numerous cellular exonucleases (which only attack free ends of DNA), the new DNA must either self-circularize (e.g., as plasmid DNA will) or be stabilized into the bacteria's circular chromosome by homologous recombination. **Homologous recombination** (Figure 5-1) is basically an exchange of two nearly identical pieces of DNA. **It requires an area of near homology (often several genes long)** *and* **a series of recombination proteins, most critical of which is a DNA-binding protein called RecA,** which helps align the homologous DNA.

Homologous Recombination

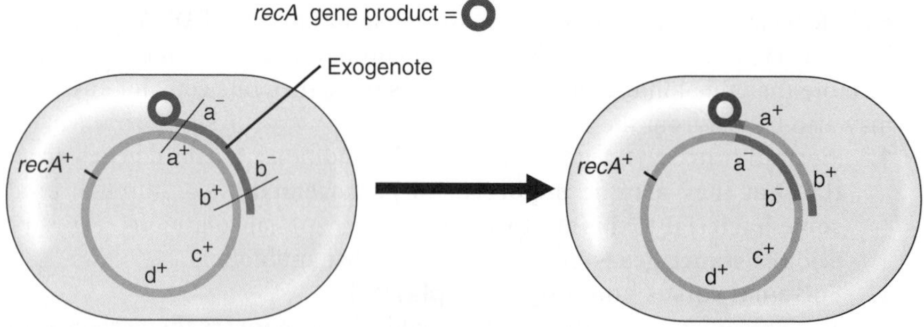

This process is required to stabilize genes introduced by transformation, transduction, and conjugation (except those that circularize). It requires a series of recombination enzymes (represented in all diagrams as the presence of the *recA* gene product).

● **Figure 5-1** Homologous recombination.

Genes recombined into the covalently closed chromosome survive; genes that end up on the linear piece are degraded. Homologous recombination stabilizes some of the bacterial genes introduced by transformation, transduction, or conjugation.

 Bacterial Gene Transfer (three processes*)

Bacterial reproduction is asexual; no new genes are introduced through reproduction, with the exception of an occasional mutation. Genetic diversity, an advantage in an ever-changing world (e.g., new antibiotics), is increased by DNA transfer, which occurs outside of reproduction. **There are three processes that transfer bacterial DNA: transformation, conjugation, and transduction.**

A. TRANSFORMATION is the binding and uptake of naked extracellular DNA by a competent, living bacterial cell (Figure 5-2). (Ⓜ Transformation = *transfer* of naked *forms* of DNA) **The recipient cell must be competent (i.e., able to bind the DNA to its surface and bring it in through the cell envelope)** for transformation to occur. Cells become competent under certain environmental conditions (which you do not need to know for USMLE Step 1).

B. CONJUGATION is the transfer of DNA directly from one living bacterium to another by direct contact. The transfer is unidirectional from the "male" donor cell to the "female" recipient. **The cells must physically touch** (they are brought together by sex pili). This process is **under the control of a series of special genes** on a fertility factor (a plasmid); these genes are **called transfer genes,** or the *tra operon*. Figure 5-3 shows important genes or regions of a typical fertility factor and what they each do. (A quick review of this figure will make the remaining figures in this chapter easier to understand.)

1. Donor cells must have fertility factors. There are two major types of donor cells:
   a. **An F^+ cell is defined as a cell with a free fertility factor** (not inserted into the bacterial chromosome of the cell).
   b. **An Hfr cell has the fertility factor inserted into the bacterial chromosome.** It has one large, covalently closed molecule of DNA containing both the chromosome and the fertility factor.
2. **Recipient cells (F^- cells) lack fertility factors;** they have only the bacterial chromosome. **The recipient in every cross is an F^- cell.**
3. Cell types are diagrammed in Figure 5-4.
4. Two important crosses are
   a. **$F^+ \times F^-$.** Only a single strand of the plasmid DNA is transferred from the donor to the recipient; the donor genotype stays the same (F^+), but the recipient cell also becomes F^+, as detailed in Figure 5-5.
   b. **Hfr × F^-.** Because the donor cell's fertility factor is integrated into the bacterial chromosome, the fertility factor promotes the single-strand transfer of part of the fertility factor and then some of the adjoining bacterial genes (in linear order). The recipient gets some new bacterial genes but does not become an Hfr cell (no sex change), because the whole chromosome is rarely transferred (Figure 5-6).

*The definitions of *transformation, conjugation,* and *transduction* will probably get you through 50% of the bacterial genetic questions. If you want to understand more and score higher, pay attention to the figures.

Gene Transfer: Bacterial Transformation

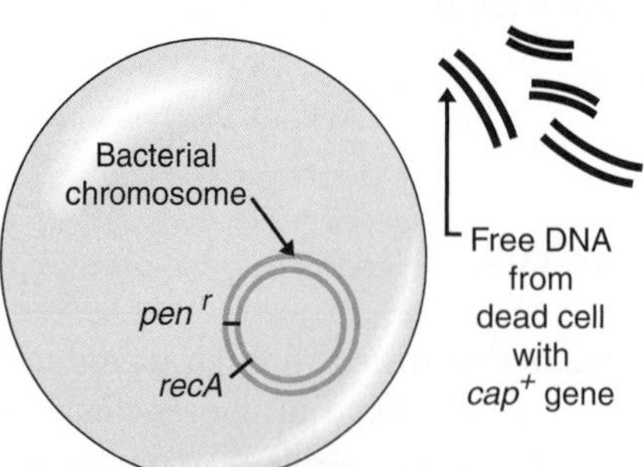

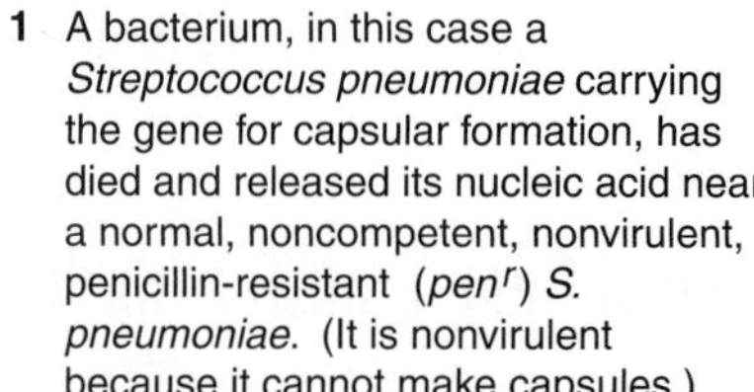

1 A bacterium, in this case a *Streptococcus pneumoniae* carrying the gene for capsular formation, has died and released its nucleic acid near a normal, noncompetent, nonvirulent, penicillin-resistant (*pen*^{r}) *S. pneumoniae.* (It is nonvirulent because it cannot make capsules.)

Most bacteria do not bind and take up DNA, but under certain growth conditions, the ability to do this (called *competency*) occurs.
Competent cells can bind and take up DNA. Competency is required for transformation.

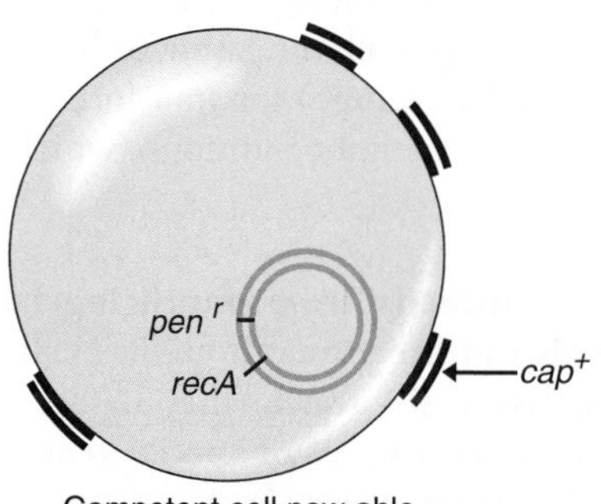

2 The extracellular free DNA binds to the competent cell and is taken up. (Details like single-strand take-up are not important.)

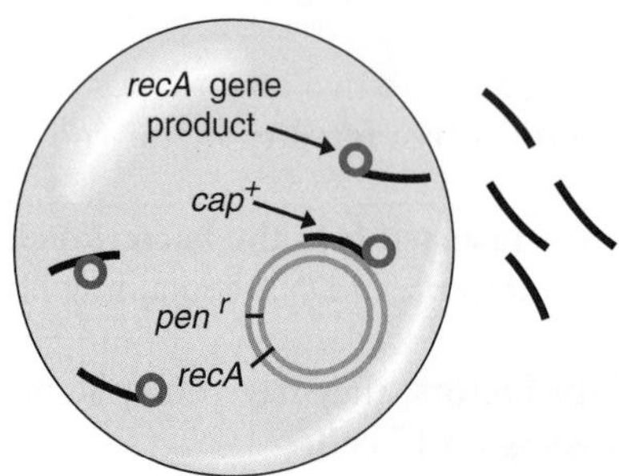

3 As long as the cell has a functioning recombination system (represented by the circle labeled *recA* gene product), each DNA can find its area of near homology, and **homologous recombination** may mediate the exchange of nearly homologous pieces of DNA.

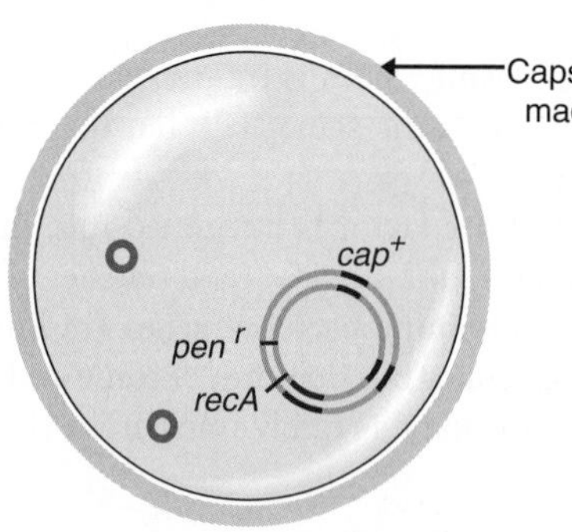

4 Stabilization of some genes by homologous recombination has occurred, and the cell has duplicated. The cell is now permanently able to make capsules.

● **Figure 5-2** Transformation is one of three bacterial DNA transfer processes increasing genetic diversity.

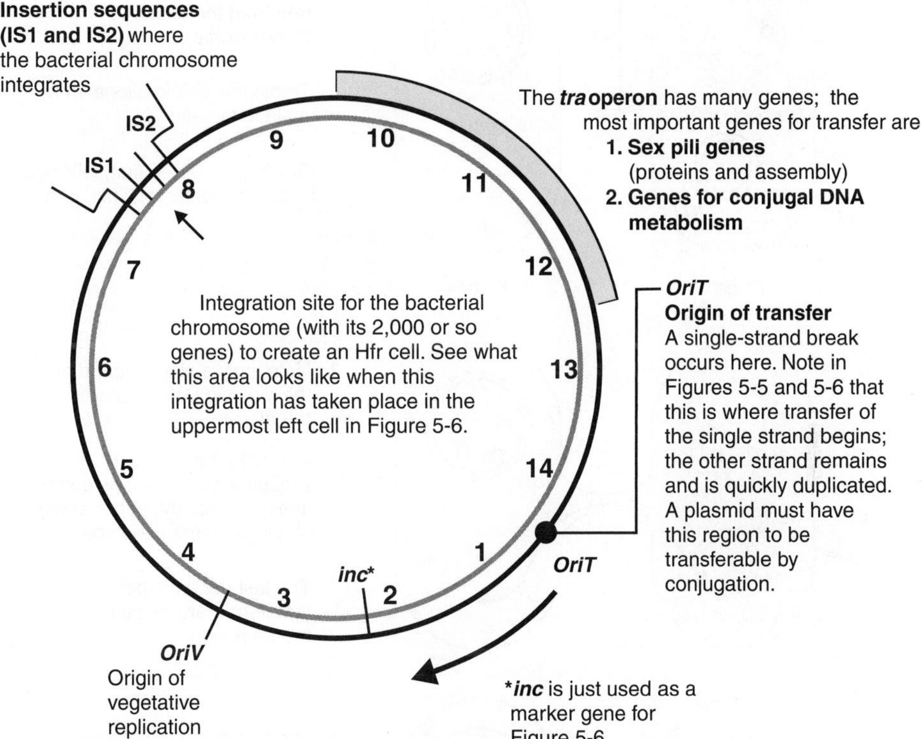

F Factor (a.k.a. F plasmid or fertility factor)
Numbers represent order of conjugal transfer in approximately 20-second blocks. Start at OriT and study diagram clockwise.
Insertion sequences (IS1 and IS2) where the bacterial chromosome integrates
IS2
IS1
1
2
3
4
5
6
7
8
9
10
11
12
13
14
The tra operon has many genes; the most important genes for transfer are
1. Sex pili genes (proteins and assembly)
2. Genes for conjugal DNA metabolism
OriT
Origin of transfer
A single-strand break occurs here. Note in Figures 5-5 and 5-6 that this is where transfer of the single strand begins; the other strand remains and is quickly duplicated. A plasmid must have this region to be transferable by conjugation.
Integration site for the bacterial chromosome (with its 2,000 or so genes) to create an Hfr cell. See what this area looks like when this integration has taken place in the uppermost left cell in Figure 5-6.
inc*
OriT
OriV
Origin of vegetative replication
*inc is just used as a marker gene for Figure 5-6.

● **Figure 5-3** Conjugation: Important genes and regions of a typical fertility factor.

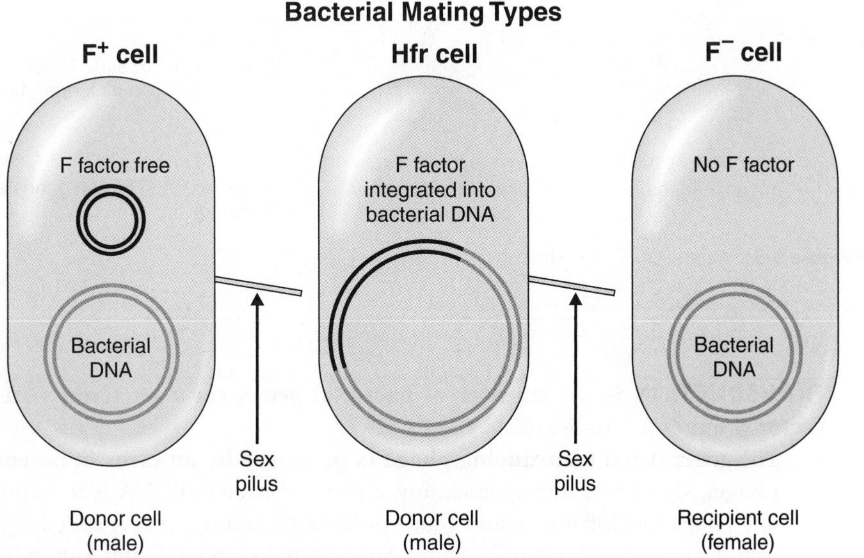

● **Figure 5-4** Conjugation: Bacterial mating types.

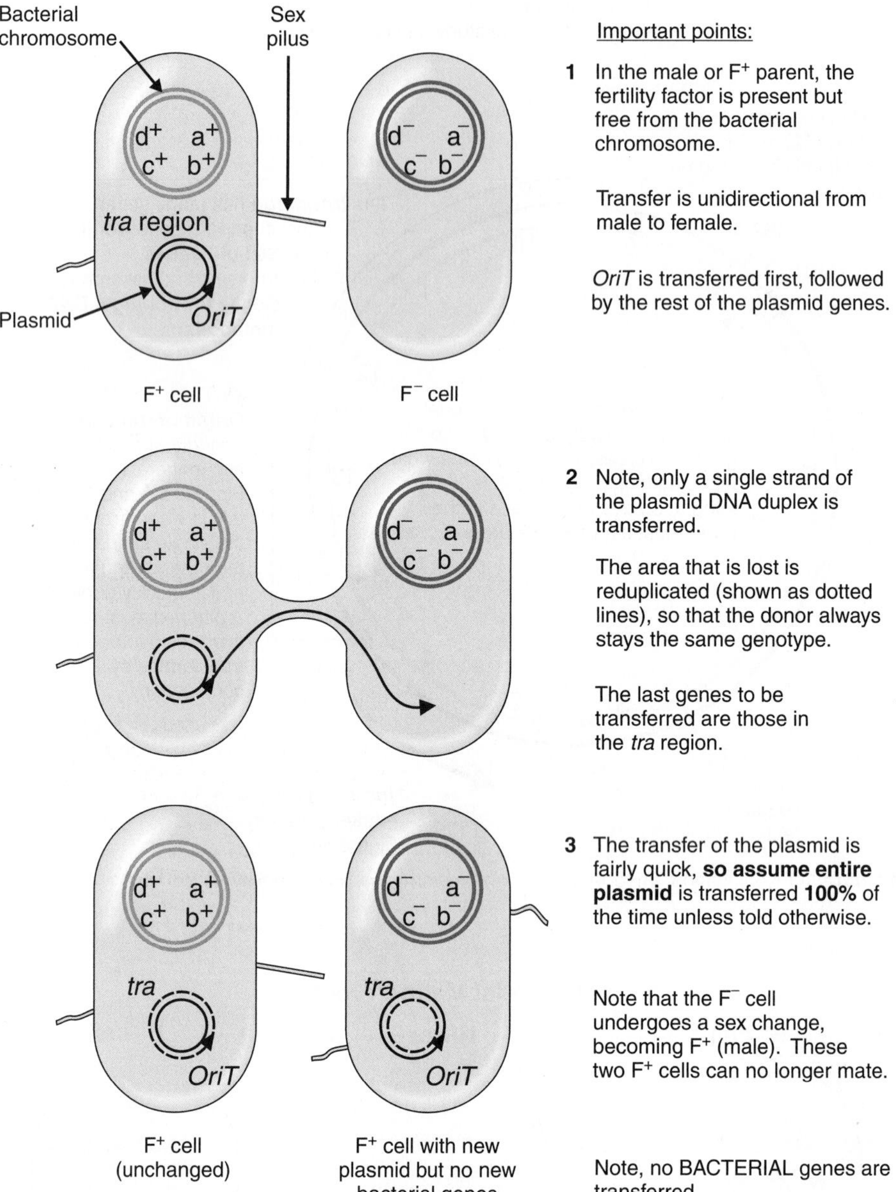

Bacterial Conjugation: F+ × F− Mating
Bacterial chromosome
Sex pilus
d+ a+
c+ b+
tra region
Plasmid
OriT
d− a−
c− b−
F+ cell
F− cell
Important points:
1 In the male or F+ parent, the fertility factor is present but free from the bacterial chromosome.
Transfer is unidirectional from male to female.
OriT is transferred first, followed by the rest of the plasmid genes.
2 Note, only a single strand of the plasmid DNA duplex is transferred.
The area that is lost is reduplicated (shown as dotted lines), so that the donor always stays the same genotype.
The last genes to be transferred are those in the tra region.
3 The transfer of the plasmid is fairly quick, so assume entire plasmid is transferred 100% of the time unless told otherwise.
tra
Note that the F− cell undergoes a sex change, becoming F+ (male). These two F+ cells can no longer mate.
F+ cell (unchanged)
F+ cell with new plasmid but no new bacterial genes
Note, no BACTERIAL genes are transferred.

● **Figure 5-5** Conjugation: $F^+ \times F^-$ cross.

C. **TRANSDUCTION is the transfer of bacterial genes via a bacterial virus (phage) vector** (Figures 5-7 to 5-11).

1. The **generalized transducing phage is produced by an error in assembling the phage.** Sometimes during assembly, a piece of bacterial DNA is incorporated into one phage head. Because any gene could be picked up, this is called a *generalized transducing phage* (assembly error). See Figure 5-7, Step 7, and Figure 5-8.

Bacterial Conjugation: Hfr × F⁻ Cross

Newly synthesized DNA is shown as dashed lines.

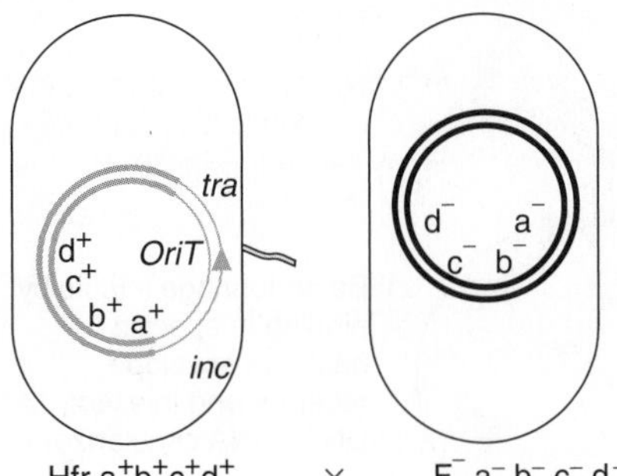

Important points:

1 Hfr donor means that the fertility factor (fine line) is already integrated into the bacterial chromosome (heavier gray line).

In this cross, plasmid genes starting at *OriT* will be transferred first, followed by the bacterial genes in linear order away from the plasmid.

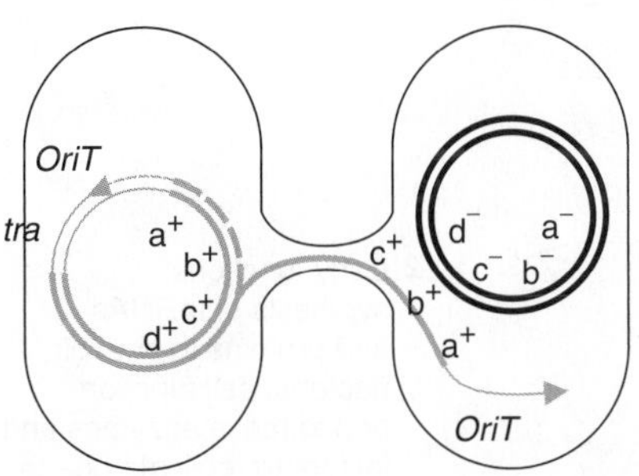

2 Note that as with the $F^+ \times F^-$ cross, only a single strand of the DNA duplex is transferred. The area that is transferred is reduplicated (note the rolling model at left), so that the donor always stays the same genotype.

IF the entire chromosome were to be transferred, the last genes to be transferred would be the *tra* region.

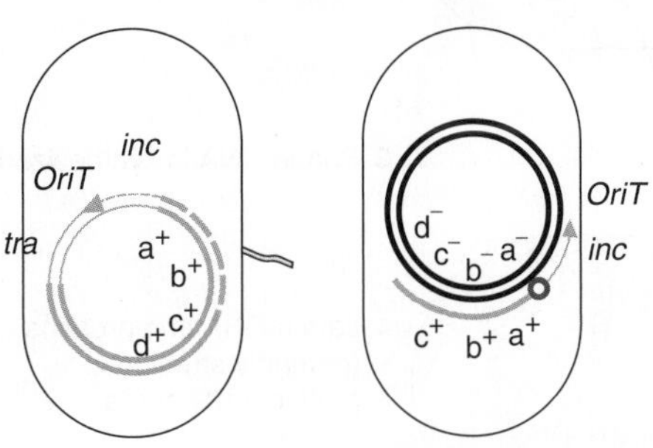

3 It takes approximately 2 hours for a complete transfer to occur. Because the cytoplasmic bridge and DNA are so fine, mating is normally interrupted before the transfer is complete. Assume that mating is interrupted and the recipient gets some new genes but (because it does not get the *tra* operon) does not become Hfr.

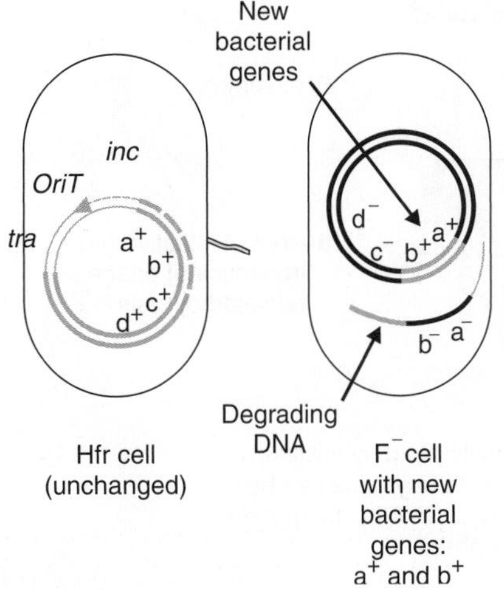

● **Figure 5-6** Conjugation: Hfr × F⁻ cross.

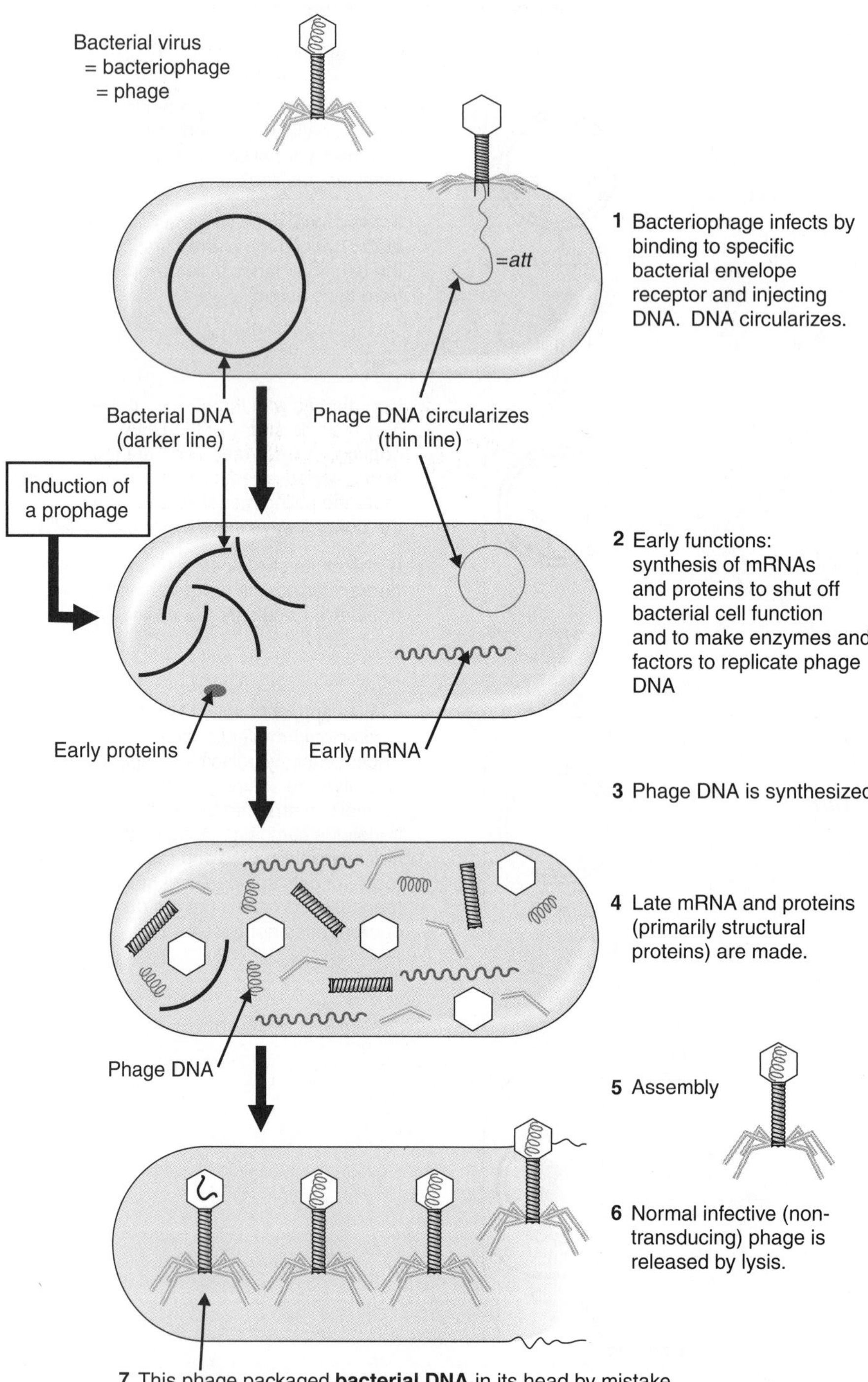

Lytic Replication of Phage
Bacterial virus
= bacteriophage
= phage
=att
1 Bacteriophage infects by binding to specific bacterial envelope receptor and injecting DNA. DNA circularizes.
Bacterial DNA
(darker line)
Phage DNA circularizes
(thin line)
Induction of
a prophage
2 Early functions: synthesis of mRNAs and proteins to shut off bacterial cell function and to make enzymes and factors to replicate phage DNA
Early proteins
Early mRNA
3 Phage DNA is synthesized.
4 Late mRNA and proteins (primarily structural proteins) are made.
Phage DNA
5 Assembly
6 Normal infective (non-transducing) phage is released by lysis.
7 This phage packaged bacterial DNA in its head by mistake. It is called a transducing phage. Because any gene can be incorporated (depending on what bacterial DNA is incorporated), it is called a generalized transducing phage.

● **Figure 5-7** Transduction: Lytic replication of phage.

Generalized Transduction

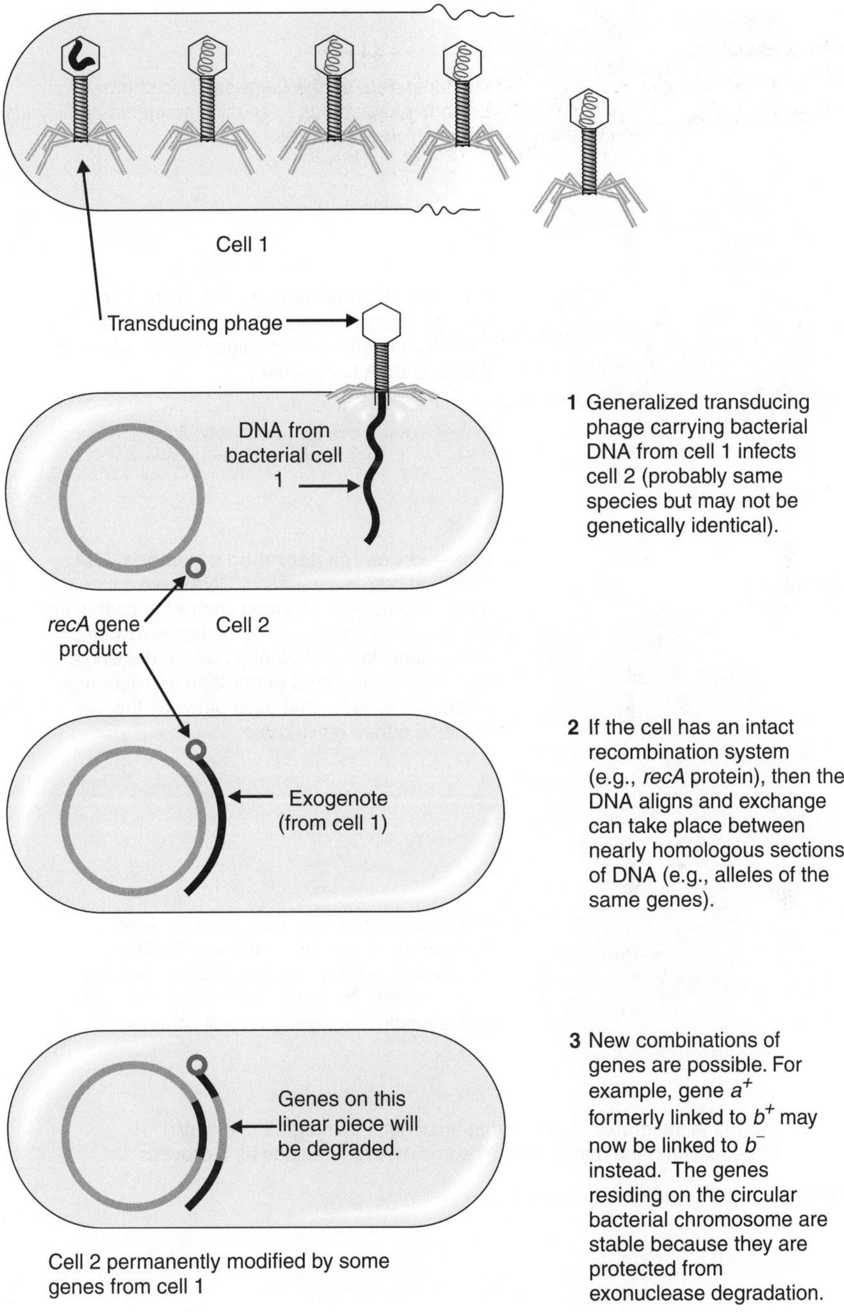

Cell 1
Transducing phage
DNA from bacterial cell 1
1 Generalized transducing phage carrying bacterial DNA from cell 1 infects cell 2 (probably same species but may not be genetically identical).
recA gene product
Cell 2
Exogenote (from cell 1)
2 If the cell has an intact recombination system (e.g., recA protein), then the DNA aligns and exchange can take place between nearly homologous sections of DNA (e.g., alleles of the same genes).
Genes on this linear piece will be degraded.
3 New combinations of genes are possible. For example, gene a+ formerly linked to b+ may now be linked to b− instead. The genes residing on the circular bacterial chromosome are stable because they are protected from exonuclease degradation.
Cell 2 permanently modified by some genes from cell 1

In generalized transduction, every bacterial gene has an equal chance of being incorporated into the phage head and being transferred to the next bacterial cell that is infected.

● **Figure 5-8** Generalized transduction.

Temperate Phage and Lysogeny

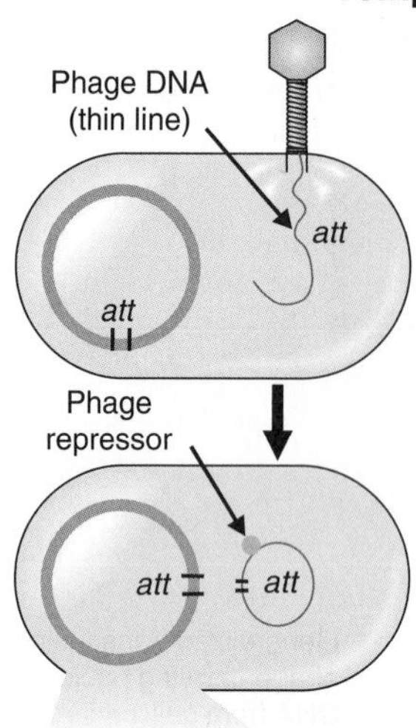

1 The temperate phage Lambda (λ) is shown. Lambda phage binds to specific receptors and injects DNA, which circularizes.

2 If functional repressor protein is made quickly enough, it inhibits transcription of structural proteins and active production of virus, allowing the virus DNA to integrate.

Phage could have gone into lytic life cycle here if the regulatory battles had gone differently.

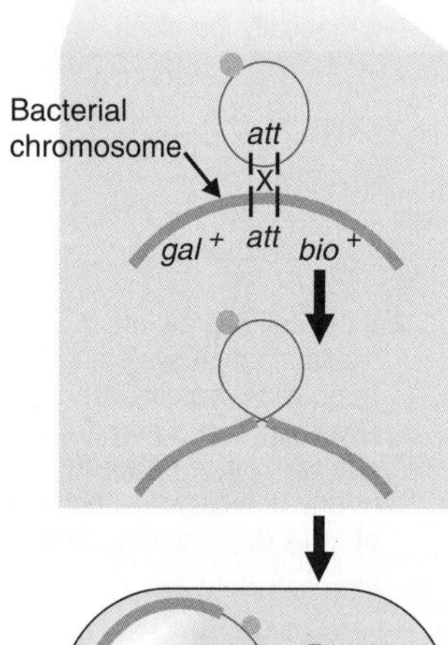

3 Enlarged view of **integration of lambda DNA:** Note that both molecules of DNA have a small area of homology (*att* sites) where the pairing and crossing over occur. This is a classic example of site-specific recombination where the whole molecule is integrated rather than an exchange taking place. Note that *att* is between the bacterial genes *gal* and *bio*.

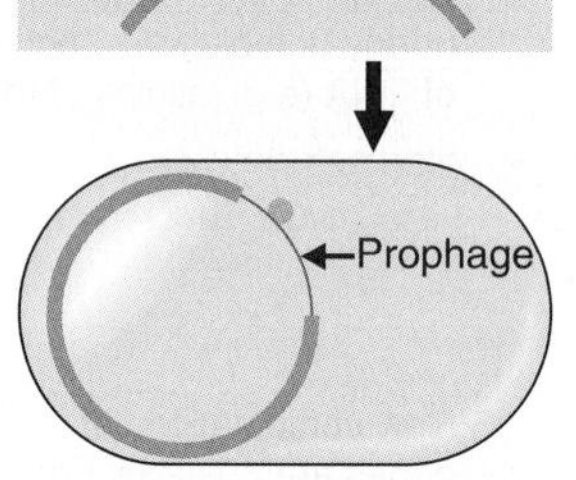

4 This is a **lysogenized cell** (state is called **lysogeny**). When the host bacterial DNA duplicates, so does the phage DNA. As long as the repressor protein continues to be made and is functioning, lysogeny will continue.

5 Prophage integration is somewhat analogous to integration of HIV DNA copy into the human chromosome, where it resides as a provirus.

● **Figure 5-9** Transduction: Temperate phage and lysogeny.

2. **Specialized transducing phage is created when there is an error in the excision of a prophage.** Temperate phage inserts only at specific sites on the bacterial chromosome. See Figure 5-9. Therefore, when the bacterial DNA is picked up as an integrated prophage is being excised (Figure 5-10 right side), only genes adjoining the integration site can be picked up—thus, the name *specialized* or *restricted transducing phage* (excisional error; Figures 5-10 and 5-11).
3. For USMLE Step 1, learn the definitions of the two types of transduction. This information is depicted graphically in Figures 5-7 through 5-11.

Induction/excision of prophage leads to a. active temperate phage replication or b. production of specialized transducing phage

If the repressor in a lysogenized cell is damaged by ultraviolet light, cold, or alkylating agents, the cell is "induced" into active virus production, which begins with the excision of the prophage DNA. Excision is the reverse of site-specific integration. This is shown below on the left with the normal process and on the right with an occasional error leading to production of the specialized transducing virus.

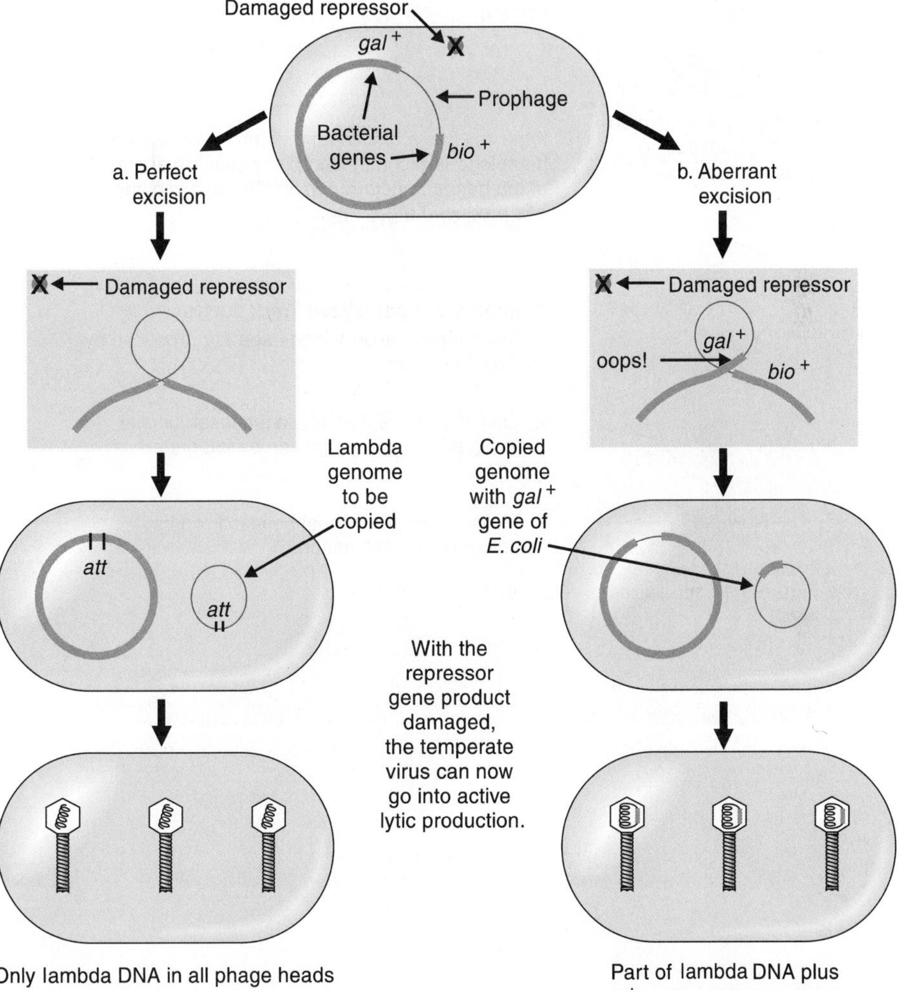

Normal lambda phage are generally produced. These are not transducing. If after a perfect DNA excision there is a late DNA packaging error, then a generalized transducing phage could be produced along with these normal nontransducing phage.

Specialized lambda transducing phage are produced, each of which is carrying the *gal*$^+$ gene. (Only the *gal* **or** *bio* genes could have been picked up.)

● **Figure 5-10** Transduction: Induction/excision of prophage.

Specialized Transduction = Restricted Transduction

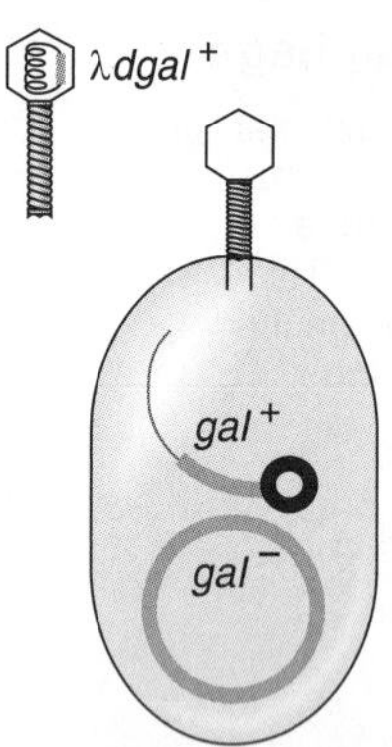

E. coli gal − cell infected with a transducing phage carrying gal +

1 Transducing phage created in a *gal* + cell by induction of a prophage. This defective phage has some lambda genes and the bacterial gene *gal* +.

2 Transducing *gal* + phage has now injected its DNA into a *gal* − cell.

3 If the cell has a functional recombination system, homologous recombination may stabilize some of the genes transferred in. In this case, recombination has produced a *gal* + cell.

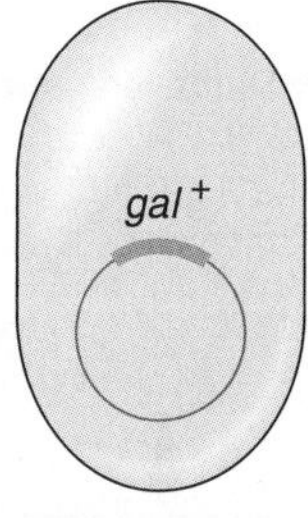

Now *gal* +

Summary of Specialized Transduction

1. Specialized transducing phage are produced by an excisional error.

2. Only the genes that adjoin the insertion site (*att*) of a temperate phage can be integrated into the phage.

3. Transduced genes must be stabilized by homologous recombination.

● **Figure 5-11** Specialized or restricted transduction.

Chapter 6

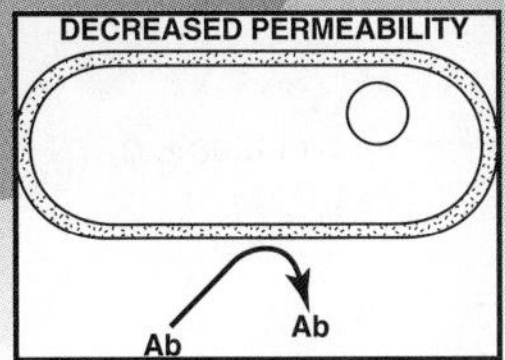

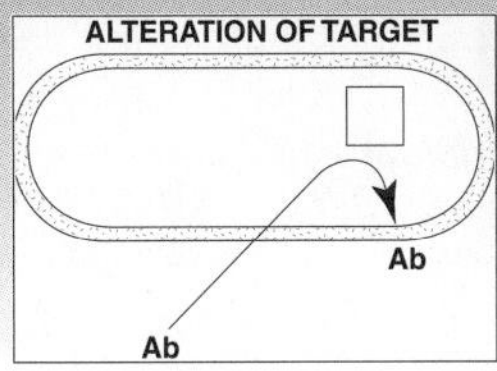

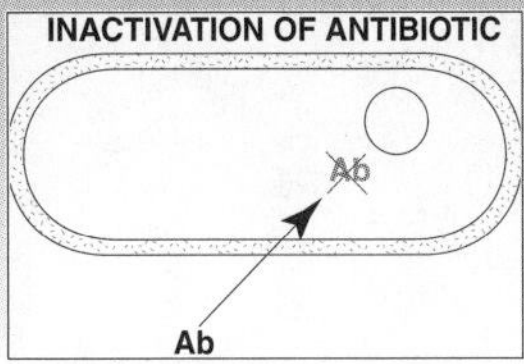

Antibiotics and Drug Resistance

I Antibiotic Mechanisms

Antibiotics inhibit bacterial processes or damage bacterial cells. **Major antibiotics and their mechanisms** are listed in Table 6-1. (See also *High-Yield Pharmacology,* 2nd ed. Philadelphia, PA: Lippincott Williams & Wilkins; 2003.) These mechansims are summarized graphically in Figure 6-1.

II Antibiotic Effectiveness

Antibiotic effectiveness depends on the properties of the antibiotic, the type and state of the bacterial cell, the microniche of the bacterium (pH, presence of fibrin, whether it is intracellular or extracellular), and whether the antibiotic penetrates the site (which may be modified by inflammation.)

A. **BACTERIO*STATIC* ANTIBIOTICS *INHIBIT* BACTERIAL GROWTH** while the drug is present.

B. **BACTERIO*CIDAL* ANTIBIOTICS *KILL* BACTERIAL CELLS.** Cells do not regrow if the antibiotic is removed.

C. **A BACTERIOCIDAL DRUG may be bacteriostatic under certain conditions,** including reduced metabolic rates in biofilms or lowered pH from necrosis.

D. **WHILE THE IMMUNE SYSTEM OF IMMUNO*COMPETENT* PEOPLE** will aid **a bacteriostatic antimicrobial** to clear up an infection, the immune system of an immunocompromised (IC) person may not. IC patients often require continuing maintenance therapy to prevent relapse.

E. **ACTIVITY MAY BE CONCENTRATION-DEPENDENT (aminoglycosides, fluoroquinolones, and metronidazole)**—the higher the concentration, the faster the killing. By contrast, maintaining concentrations at or close to the minimal inhibitory concentration (MIC) for treatment time is critical (time above MIC) **for beta lactams, macrolides, vancomycin, and clindamycin.**

TABLE 6-1 ANTIBIOTIC MECHANISMS AND DRUG RESISTANCE

	Mechanism of Antibiotic Action	Antibiotics	Mechanisms of Bacterial Resistance
Nucleic acid synthesis inhibition	Nucleic acid precursor: folic acid synthesis inhibition	Sulfonamides/trimethroprim	Altered enzymes; reduced permeability or increased efflux
	Inhibition of mRNA synthesis (DNA-dependent RNA polymerase)	Rifampin	Reduced binding to DNA polymerase
	Inhibition of DNA synthesis DNA gyrase inhibition	Fluoroquinolones	Altered gyrase, reduced permeability
	Electron sink; intermediates damage DNA (anaerobes only)	Metronidazole	DR is uncommon in anaerobes; drug not converted to active form in aerobes.
Inhibition of protein synthesis and assembly	Binds 50S ribosomal subunit: blocks translocation	Macrolides Clindamycin	Methylating enzyme Methylating enzyme
	Binds 50S ribosomal subunit	Linezolid	?
	Binds 50S: Inhibits chain elongation	Streptogramins (quinupristin-dalfopristin)	
	Binds 50S: Blocks peptidyl transferase	Chloramphenicol	Acetyltransferase
	Binds 30S: blocks tRNA attachment	Tetracycline	Increased efflux
	Binds 30 and 50S: disrupts initiation complexes	Aminoglycosides (aerobes that can transport)	Decreased ribosomal binding; reduced uptake; modifying enzymes
Cell membrane damage	Membrane (outer and cytoplasmic) disrupted	Polymyxins	Colistin
Cell wall synthesis inhibition	Early steps in peptidoglycan synthesis	Cycloserine, bacitracin	New, insensitive peptidoglycan precursors
		Vancomycin	Precursor d-ala-d-ala is converted to d-lactate-d-lactate, which is insensitive to vancomycin.
	Cross-linking of peptidoglycan (transpeptidation)	Penicillins/cephalosporins, imipenem, aztreonam	Altered PBPs; reduced permeability; beta-lactamase
	Mycolic acid synthesis	Isoniazid	Loss of catalase function by mycobacterium leads to failure to activate prodrug.

KEY: PBPs, penicillin-binding proteins.

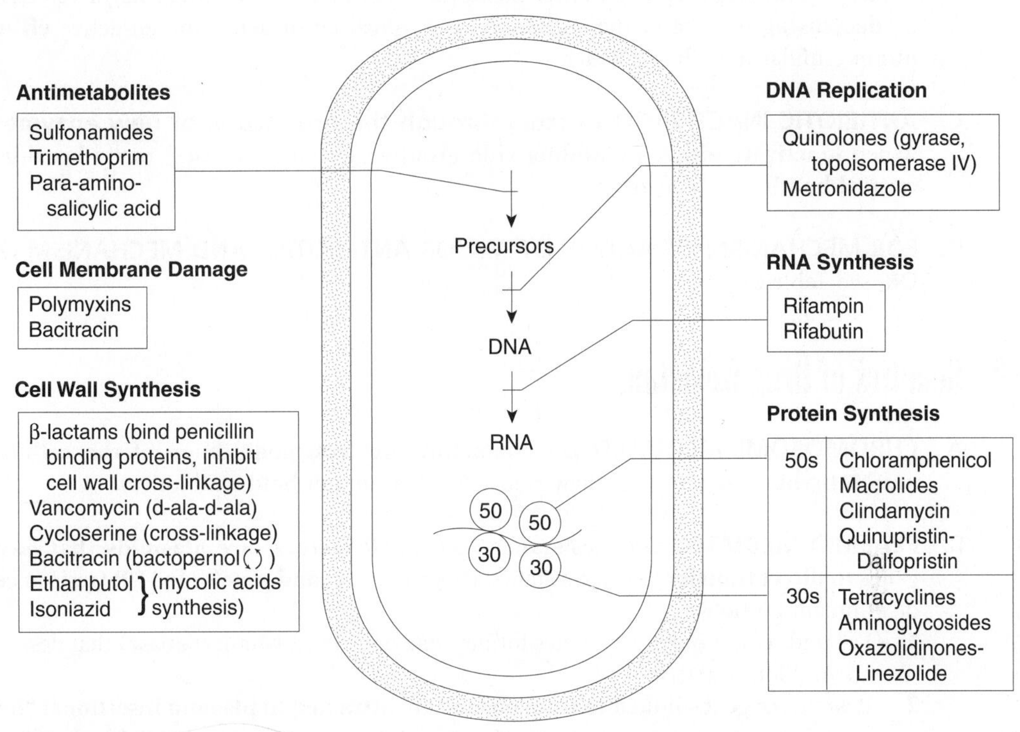

● **Figure 6-1** Antibiotic targets.

 Combination Drug Therapy

Specific combinations of antibiotics are used (1) to broaden the spectrum for polymicrobic therapy, (2) for initial empiric therapy until agents and susceptibilities are known, or (3) to prevent or overcome resistance (beta-lactam with beta-lactamase inhibitor). Limited situations may require combinations, such as beta-lactam to weaken the *Enterococcus* cell wall so that the aminoglycoside can enter the cell. This is an example of synergism—the effect of the two is greater than either alone. (Antagonism is when the combination is worse than the two alone.)

 Drug-Resistance Mechanisms

Bacteria-producing antibiotics have to be resistant to that antibiotic. Thus, resistance to natural antibiotics precedes commercial use. Some organisms may have inherent drug resistance (DR) because the process or structure is not inhibited. A variety of alterations in bacterial cell structure or functions or the production of inactivating enzymes lead to DR.

A. MODIFICATION OF THE DRUG'S BINDING SITE allows normal cell function to continue, even in the presence of the drug. Examples are new penicillin-binding proteins (PBPs) or new ribosome subunits to which the drugs will no longer bind.

B. **REDUCED ACCESS OF THE DRUG TO ITS ACTIVE SITE** may occur by **decreasing bacterial permeability** (e.g., outer membrane modifications of Gram-negative cells), by **decreasing uptake** of the antibiotic, or by creating or activating an active **efflux pump** (similar to bailing a boat).

C. **ANTIBIOTIC INACTIVATION occurs through the production of new enzymes,** which **inactivate either by adding side groups** (acetyltransferases) or by **breaking critical bonds** (beta-lactamases).

D. **FOR MECHANISM OF ACTION OF MAJOR ANTIBIOTICS AND MECHANISM OF DR,** see Table 6-1.

V Genetics of Drug Resistance

A. **CHROMOSOME-MEDIATED DR** commonly involves modification of **the cellular antibiotic-binding site** (e.g., new ribosomes that do not bind antibiotics).

B. **PLASMID-MEDIATED DR. Resistance factors (R factors) are plasmids that have genes to direct transfer** (a.k.a. resistance transfer factor) **and gene(s) for DR** (resistance-determinant portion).

1. Plasmids commonly carry genes for new enzymes (e.g., beta-lactamase) that destroy an antibiotic's activity.
2. **Resistance genes linked to transposons are attracted to plasmid insertional "hot spots"** (insertion sequences), **creating multiple drug resistance (MDR) plasmids.**
3. Plasmid genes are easily transmitted by conjugation.
   a. Rapid transfer of **MDR** to another cell is most likely conjugal transfer of MDR plasmid.
   b. **Nonconjugative plasmids** have lost their transfer (*tra*) region but still **may be transferred by conjugation in a process called *mobilization*,** as long as there is another conjugative plasmid in the cell. (The USMLE loves this paradox!) The most notable example is *Neisseria gonorrhoeae*.

C. **SOME RESISTANCE GENES** are regulated so that their product is only produced in the presence of the antibiotic. For example, although many beta-lactamases are constitutively produced, others are turned on only by exposure to the antibiotic.

VI Acquisition of Drug Resistance

A. **ALREADY DR STRAINS SPREAD**

1. In nosocomial infections, hospital staff may serve as reservoirs for methicillin-resistant *Staphylococcus aureus* (MRSA) and, by improper hygiene, may transmit MRSA to patients. Poor medical staff hygiene spreads microbes from one patient to another.
2. MDR *N. gonorrhoeae* and *Mycobacterium tuberculosis* can spread by contact within the community.

B. **DR GENES may be transferred from normal flora to a newly acquired pathogen.** This is most likely with Enterobacteriaceae in gastrointestinal infections, but it is also a concern with skin and mucosal flora.

C. **RESISTANCE AND SELECTION may occur through mutation,** most notably with fluoroquinolones or when a patient uses no drug or one, instead of two or three, during the long antibacterial treatment for tuberculosis.

D. **DAMAGE BY A VIRULENT ORGANISM** may lead to changed environmental conditions at the site of infection and reduced potency of an antibiotic (e.g., decreased pH in abscesses lowers effectiveness of aminoglycosides.).

VII Bacteria with Major Drug Resistance

A. GRAM-POSITIVE BACTERIA

1. ***Enterococcus.*** Some strains show resistance to vancomycin and streptomycin/gentamicin.
2. ***Staphylococcus aureus.*** Some **MRSA** (methicillin resistant *S. aureus*) strains are resistant to all drugs in common usage, except vancomycin. (Although some now also have intermediate resistance to vancomycin.) Methicillin resistance is due to a chromosomal modification of a major PBP; other resistance is due to an MDR plasmid transferred by transduction.
3. ***Streptococcus pneumoniae.*** Both low- and high-level penicillin resistance is increasing as a result of **mutations in PBP genes.**

B. GRAM-NEGATIVE BACTERIA

1. **Enterobacteriaceae.** MDR plasmids are transferred by **conjugation.** An example of the promiscuity of bacteria is that MDR plasmids can be transferred from *Escherichia coli* to related pathogens (e.g., *Salmonella*).
2. ***Neisseria gonorrhoeae.*** The beta-lactamase gene resides on nonconjugative plasmids, transferred by mobilization (conjugation.)
3. ***Haemophilus influenzae.*** This bacterium has resistance to many antibiotics.
4. ***Pseudomonas.*** This opportunist has inherent resistance (missing some high-affinity porins through which some drugs enter) and has picked up some plasmid-mediated DR, which makes it tough to treat in cystic fibrosis patients (microcolonies) and IC patients.

C. NON-GRAM-STAINING BACTERIA

1. ***Mycobacterium tuberculosis.*** **MDR strains are increasingly common.** A patient may acquire an MDR strain or may start with a drug-sensitive strain and, through improper adherence to a drug protocol (i.e., sporadic compliance or taking too few drugs at a time), may select for DR mutants. Resistance to drugs that bind to ribosomes is particularly common because ***M. tuberculosis*** (unlike most bacteria) appears to have only one copy of each of the ribosomal genes.
2. ***Mycoplasma.*** This organism has **inherent DR** to all cell wall active drugs.
3. ***Rickettsias* and *chlamydiae*** are obligate intracellular organisms; therefore, they are resistant to such drugs as vancomycin, aminoglycosides, and any other drugs that do not penetrate human cells.

Chapter 7

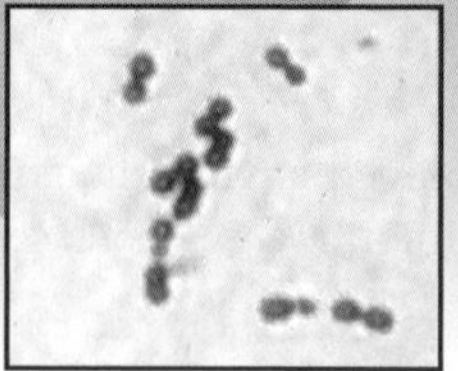
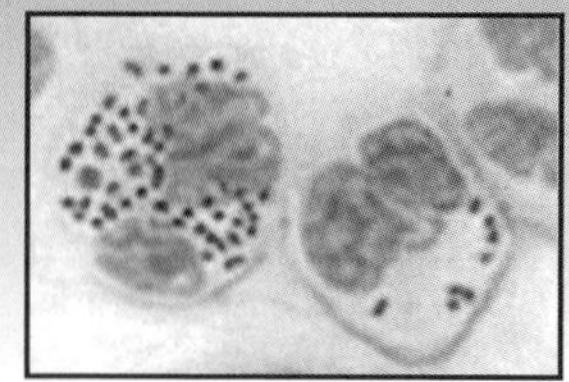
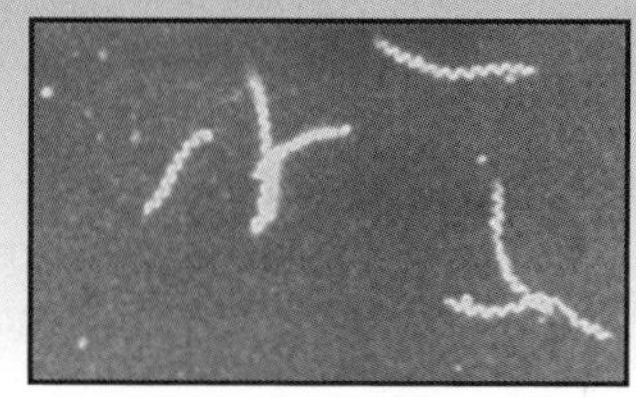

Identification of Major Bacterial Groups

Methods used to place bacteria into taxonomic groups often differ from those used to identify clinical isolates. Taxonomic classification uses primarily genotypic methods (e.g., DNA hybridization and sequencing techniques, ribotyping, analysis of plasmids, and restriction endonuclease fragment analysis) and analytic methods (e.g., cell surface composition or protein analysis). Diagnostic microbiology has relied on phenotypic tests in the past, but as genotypic methods become easier to do, they are being used more routinely in clinical labs. Particularly important is identifying the presence of genes for a toxin or a gene for drug resistance. Serologic assays to identify immune response to an agent or to demonstrate the presence of specific antigens may also be useful, especially if the agent cannot be cultured or is dangerous to culture. In addition, one of the most important functions of the lab is to determine drug susceptibilities or resistance.

The physician's assessment of the patient's potential site of infection and most likely causative agents leads to orders for specific specimens. The clinical microbiology lab then chooses the most appropriate transport media and initial tests, such as microscopy, to determine which additional lab tests will most likely identify the agent. The lab has specific media and procedures that are routinely used for certain specimens or when a particular agent is suspected. (For example, abscess specimens may be taken in a sterile syringe [with no air in it], transported quickly to the lab, and cultured anaerobically.) The lab must be notified if the physician suspects an unusual agent that requires special conditions or tests for identification.

I Clinical Laboratory Identification of Bacteria

A. TESTS DONE DIRECTLY ON PRIMARY SPECIMENS. Tests used depend on the specimen type and the most likely agents. Tests may include

1. **Microscopic visualization** (e.g., Gram stain, except where normal flora prevents interpretation; fluorescent antibody staining; dark field microscopy)
2. **Nucleic acid probe tests, with or without amplification (polymerase chain reaction)**
3. **Detection of a specific antigen or enzyme**
4. **Specimen culture**
5. **Detection of the host response (serology on serum)**

B. TESTS DONE ON ISOLATED PURE CULTURE. Once grown, the isolated pure culture may be subjected to a variety of tests, including microscopy with differential staining, biochemical assays to identify specific enzymes or metabolic pathways, tests to identify specific antigens or nucleic acid sequences, and antibiotic-susceptibility testing.

II Specificity Versus Sensitivity

Lab tests may be negative in a patient infected with the agent (false negative) or positive in a patient who does not have the agent (false positive). Certain tests may be used only at specific times in disease progression. Tests used as screening tests (e.g., in suspected syphilis), which are often inexpensive to run, must be sensitive but need not be specific. Positive screening tests that have low specificity are followed up with a more specific test, which is often more costly. See Figure 7-1 to review concepts of sensitivity and specificity.

III Specimens

Proper specimen handling is critical to successful cultural isolation. Consideration needs to be given to

A. COLLECTION METHODS. These are dependent on

1. **Site of infection.** Some sites are more likely to have anaerobes or to have a high risk of contamination with either normal flora or external contaminants.
2. **Most likely organisms to be isolated.** For instance, for *Bordetella pertussis*, special calcium alginate or Dacron nasal pharyngeal swabs on wires or cough plates are used.

B. TRANSPORT. Some organisms (e.g., staphylococci) are hardy. Others have specific transport needs. For example, because *Neisseria* are cold-intolerant, they are plated on warm chocolate or Thayer-Martin agar (chocolate with vancomycin, colistin, and nystatin) as soon as possible.

IV Cultures

A. INERT BROTHS OR AGARS are used for most bacteria. The choice depends on the specimen site and suspected agent.

1. **Common primary media.** Neither blood agar nor chocolate agar is selective. Both are enriched.
   a. Blood agar is a rich agar base supplemented with blood.
   b. Chocolate agar has lysed blood added to provide free hemin, which is required for *Neisseria* and *Haemophilus*.
2. **Specialized selective media** promote the growth of specific pathogens (or groups) and inhibit the growth of normal flora.
3. **Differential media** allow distinction by appearance between two different colonies of bacteria directly on the plate (e.g., lactose fermenters from nonfermenters on MacConkey medium).
4. **High-yield media** (likely to appear in USMLE cases) are listed in Table 7-1.

Sensitivity/Specificity

- Sensitivity measures how well a test identifies those with disease (true positives). In other words, a test with high sensitivity will not miss many with a disease, although it may identify people in the positive pool who do not actually have disease.
- Specificity refers to the ability of the test to detect only the true positives without picking up false positives. It may have lowered sensitivity and miss some true positives.

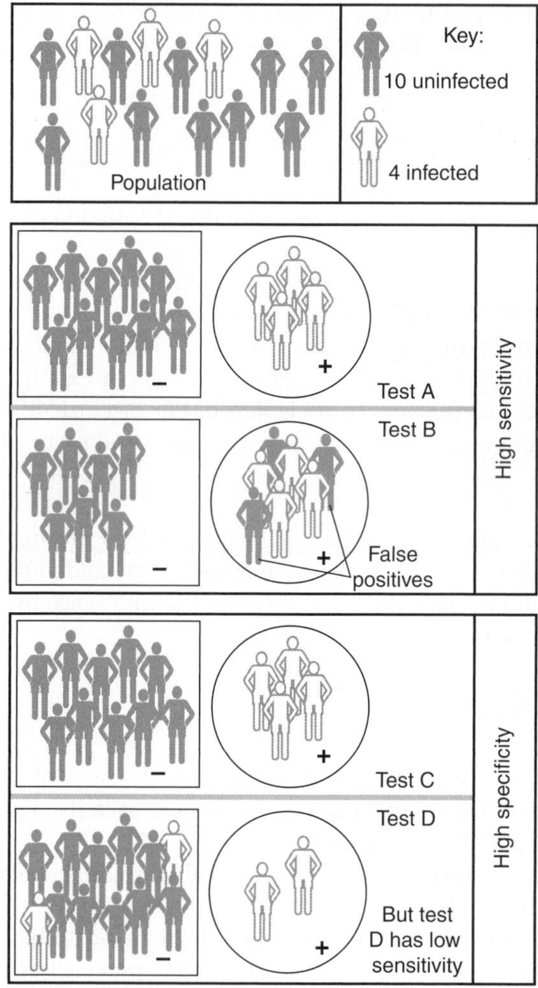

● **Figure 7-1** Laboratory test sensitivity and specificity.

TABLE 7-1 SELECTIVE AND DIFFERENTIAL BACTERIAL CULTURE MEDIA

Medium	Bacteria Isolated and/or Identified
Buffered charcoal-yeast extract (BCYE) agar	*Legionella* spp.
Chocolate agar	*Haemophilus* spp. *Neisseria* from sterile body sites
Eosin-methylene blue agar or MacConkey agar	Enteric bacteria
Hektoen enteric agar	*Salmonella, Shigella* spp.
Loeffler coagulated serum medium and tellurite medium	*Corynebacterium diphtheriae*
Löwenstein-Jensen medium	*Mycobacterium tuberculosis*
Thayer-Martin agar or New York City agar	Any *Neisseria* from body area with normal flora
Regan-Lowe agar medium	*Bordetella pertussis**
TCBS (an alkaline medium)	*Vibrio cholerae*

KEY: TCBS, thiosulfate-citrate-bile salts-sucrose. (Do not memorize long name.)
*It often is not possible to culture *Bordetella pertussis* from a vaccinated adult whose immunity has waned and who has active pertussis. It also becomes more difficult to culture from an unvaccinated person from the paroxysmal stage on. A PCR (polymerase chain reaction) test is preferable.

B. **EUKARYOTIC CELL CULTURES** must be used to culture chlamydiae, which are obligate intracellular pathogens and which will not grow in cell-free inert media.

C. **NO ROUTINE CULTURE** is available for rickettsiae or *Mycobacterium leprae* (all obligate intracellular parasites) or for *Treponema pallidum* (not intracellular).

D. **INCUBATION ATMOSPHERE AND TEMPERATURES VARY,** depending on the organism to be isolated.

V Identification of Bacteria

A. **BIOCHEMICAL TESTS** detect the presence of specific enzymes, proteins, or cell wall constituents. Table 7-2 lists some enzymes that are important because they can be identified with rapid tests or because they help distinguish major bacterial groups.

B. **ANTIGENS** can be identified in specimens or cultures using known antibodies. These tests include enzyme immunoassay (EIA or ELISA) precipitin tests, counterimmunoelectrophoresis (CIE), Western blots, and latex particle agglutination.

C. **GRAM STAIN** is an important rapid test. Many other special stains are available to aid in culture identification (e.g., acid fast and specific fluorescent antibody stains).

D. **GENE PROBES** indicate if a particular gene sequence is present (e.g., the gene for verotoxin production in an *Escherichia coli* strain), but they may require DNA amplification.

TABLE 7-2 IMPORTANT ENZYME TESTS USED IN IDENTIFICATION OF BACTERIA

Enzyme	Activity	Bacteria Tested	Test Principle
Oxidase	Cytochrome enzyme.	Gram-negative rods—aerobic or facultative anaerobic.	Chromogenic test reagent turns black.
Catalase	Hydrolyzes hydrogen peroxide to water and oxygen.	Differentiates streptococci from staphylococci. Mycobacteria speciation.	Generates O_2 bubbles.
Urease	Hydrolzes urea.	Specific bacteria, many involved in UTI and *Helicobacter pylori.*	Causes pH change when urea is broken down.
Coagulase	Fibrinogen converted to fibrin clot.	Speciates staphylococci Identifies *Yersinia pestis,* causative agent of plague.	Clots serum.

KEY: UTI, urinary tract infection.

VI Important Bacterial Genera

In the past, genus (and species) distinctions were made largely on the basis of biochemical and physical properties of bacteria (e.g., niacin production). Now molecular techniques are also used (e.g., GC content, ribotyping, and restriction length polymorphisms).

A. FIGURES 7-2 AND 7-3 show characteristics commonly used to distinguish different bacterial genera. These flowcharts point out only high-yield differences between organisms that are likely to be used as clues or to be tested directly on USMLE Step 1. Where two or

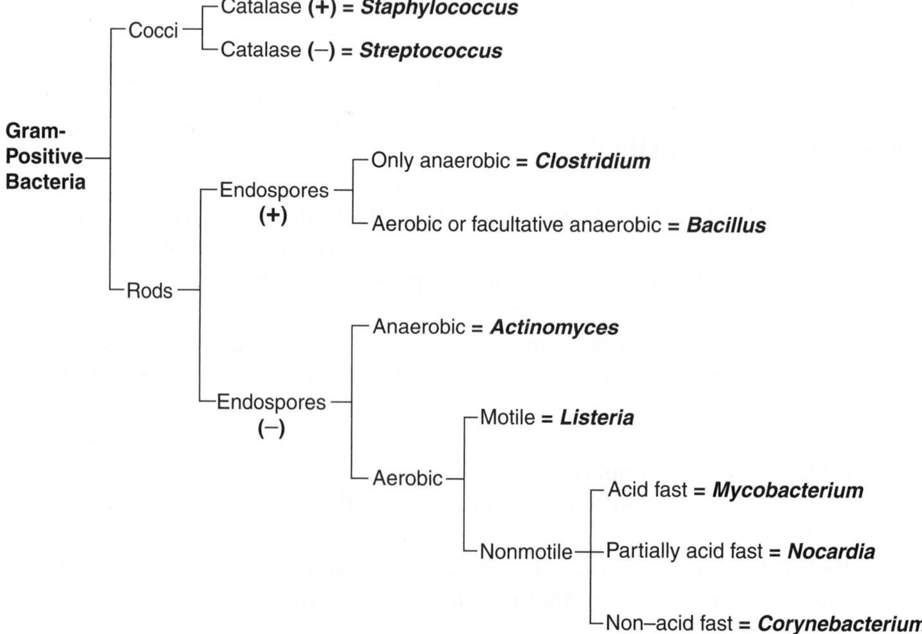

● **Figure 7-2** Flowchart for differentiating Gram-positive bacterial genera.

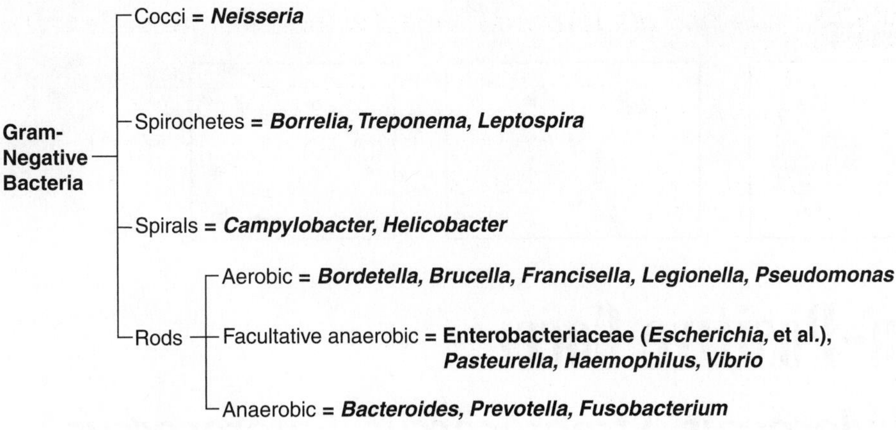

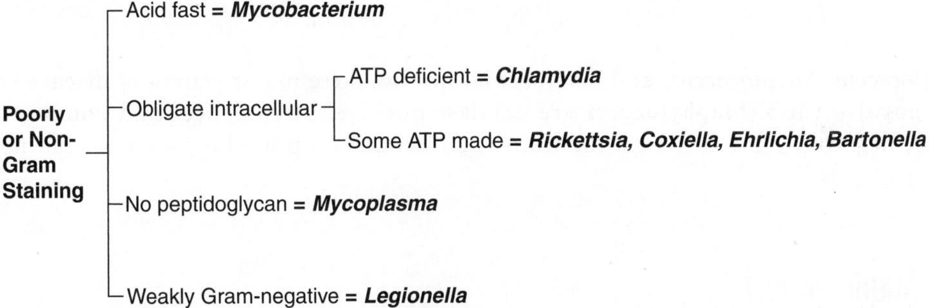

● **Figure 7-3** Flowchart for differentiating Gram-negative and poorly or non-Gram-staining bacterial genera.

more genera are listed together with no further laboratory indicators, differentiation is usually beyond the scope of USMLE Step 1.

B. NOT ALL STRAINS of a single bacterial species are identical. For example, ingested O157:H7 *E. coli* (producing verotoxin, a.k.a. shigalike toxin) may kill a child, even though other strains of *E. coli* are part of the child's normal flora. K1-encapsulated strains of *E. coli* cause neonatal meningitis.

Chapter 8

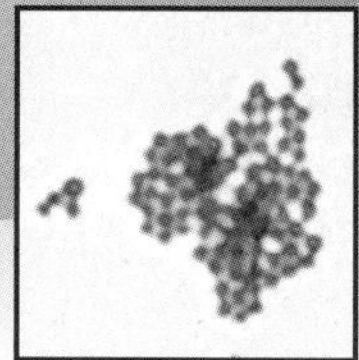
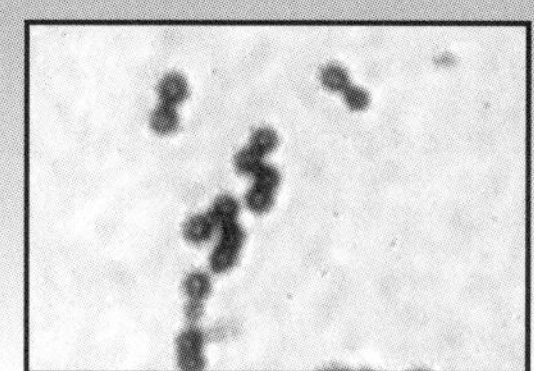
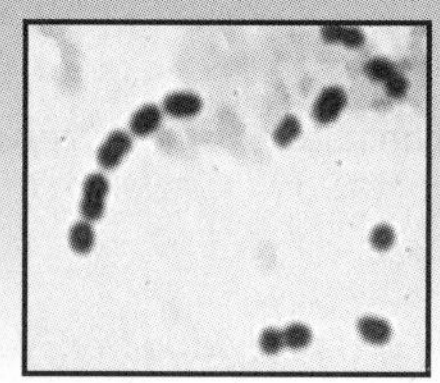

Gram-Positive Cocci

Staphylococcus, Streptococcus, Enterococcus

***Staphylococcus, Streptococcus,* and *Enterococcus* are the three major genera of disease-causing Gram-positive cocci. Staphylococci are catalase-positive. Streptococci and enterococci are catalase-negative.** ( You can get the word "c a t" from S t a p h y l o c o c c u s but not from Streptococcus.)

I Staphylococci

Staphylococci are **Gram-positive, catalase-positive cocci** that tend to **grow in clusters.** Medically important staphylococci are *S. aureus, S. epidermidis,* and *S. saprophyticus.*

A. *STAPHYLOCOCCUS AUREUS*

1. ***S. aureus*** is **coagulase-positive** (initiates formation of fibrin clot), **β-hemolytic, and salt-tolerant,** and it produces a yellow pigment. *S. aureus* has **protein A** on its surface, **which binds the Fc end of IgG-reducing opsonization and phagocytic binding.** *S. aureus* has many adhesins and enzymes and may produce exotoxins.
2. **Epidemiology. *S. aureus* colonizes nasal mucosa, skin,** and **vaginal mucosa.** Colonization is **increased in health care workers, intravenous drug abusers (IVDAs), diabetics, and neutropenic people.** It spreads via sneezing, hands, skin lesions, and fomites.
3. **Diseases**
   a. ***S. aureus* food poisoning** occurs from **heat-stable enterotoxins produced in poorly refrigerated, *S. aureus*–contaminated, high-protein foods** (e.g., canned or salted meats, custard pastries). Ingested enterotoxins are superantigens that bind vagus nerve endings, leading to rapid onset (1–6 hours) of **nausea, abdominal pain, and vomiting.**
   b. ***S. aureus*** causes **bullous impetigo** (large vesicles). *S. aureus* subcutaneous infections **(boils, mastitis, surgical wounds)** commonly present with **subcutaneous tenderness, heat, redness, pus formation, and swelling.** Infection tends to localize, but **toxins may circulate** (e.g., **exfoliation, causing scalded skin syndrome**).

c. **Toxic shock syndrome (TSS).** Surgical packing or hyperabsorbant tampons stimulate *S. aureus* growth, which may produce **TSST-1, a pyrogenic (exo)toxin superantigen (PTSAg). TSST-1** circulates, **binds major histocompatibility complex (MHC) II** (without processing antigen), and **activates many T cells, leading to massive cytokine release. Symptoms are high fever and scarlatiniform rash, followed by hypotension, vomiting, diarrhea, desquamation (particularly on palms and soles), and, eventually, multiorgan failure.**

d. **Endocarditis. *S. aureus* is the dominant cause** of bacterial endocarditis, **including in IVDAs.** The **unusual clumping factor** of *S. aureus* **allows binding to fibrinogen on normal heart endothelium.** Its **fibronectin-binding protein** enhances adhesion to a previously damaged heart. Once adhered, ***S. aureus* alpha-toxin (a pore-forming cytolysin) rapidly damages the heart.** Symptoms include **fever, malaise, leukocytosis, and development or worsening of a heart murmur.**

e. **Osteomyelitis. *S. aureus* is the major cause** of osteomyelitis in the absence of trauma or sickle cell disease.

f. ***S. aureus* pneumonia** occurs early in **cystic fibrosis** patients or in **IVDA.**

4. **Lab identification. *S. aureus*** is **Gram-positive, β-hemolytic, and catalase- and coagulase-positive. Only *S. aureus* grows on mannitol salt medium**, fermenting the mannitol. Phage typing or pulsed field gel electrophoresis differentiates strains.

5. **Drug resistance. Methicillin-resistant *S. aureus* (MRSA) contain a modified major chromosomal penicillin-binding protein (PBP).** Most MRSA strains also have plasmid-mediated resistance to all other drugs, except glycopeptides (vancomycin). **Partially vancomycin–resistant *S. aureus*** strains have been isolated. ***S. aureus* drug resistance is transferred by phage (transduction).**

B. *STAPHYLOCOCCUS EPIDERMIDIS* (the major coagulase-negative staphylococcus)

1. ***S. epidermidis*** is **coagulase-negative, nonhemolytic, and novobiocin-sensitive. It is normal skin flora.**

2. **Diseases. *S. epidermidis*** causes **catheter or prosthetic device infections**, adhering through production of **polysaccharide slime**, which leads to biofilm formation. ***S. epidermidis* endocarditis** almost always occurs in people with indwelling IV catheters or in IVDA. *S. aureus* is a far more frequent cause of endocarditis in IVDA. Drug resistance is a problem.

C. *STAPHYLOCOCCUS SAPROPHYTICUS*

1. ***S. saprophyticus*** is **coagulase-negative, nonhemolytic, novobiocin-resistant,** and **normal vaginal flora.**

2. **Diseases.** *S. saprophyticus* is one cause of urinary tract infections **(UTIs) in adolescent females, often newly sexually active.** *S. saprophyticus* UTIs are much less common than *Escherichia coli* UTIs in all sexually active women; unlike *E. coli, S. saprophyticus* is nitrite-negative.

II Streptococci

Streptococci are Gram-positive cocci arranged in pairs or chains. All are **catalase-negative, facultative anaerobes that ferment even in the presence of oxygen. Lancefield antibodies to cell wall carbohydrates** identify **serogroups** (A-U). Antibodies to capsular antigens of pneumococcus or M antigens of *Strep. pyogenes* identify serotypes. See Table 8-1.

TABLE 8-1 HIGH-YIELD STREPTOCOCCI AND ENTEROCOCCI

Species	Lancefield Group	Typical Hemolysis	Important Identifiers
Streptococcus pyogenes	A	Beta	Bacitracin-sensitive
Strep. agalactiae	B	Beta	Bacitracin-resistant; hippurate-utilized; incomplete hemolysin (CAMP factor)
Enterococcus faecalis	D	Alpha, beta, or none	Growth in 6.5% NaCl
Strep. bovis	D	Alpha or none	No growth in 6.5% NaCl
Strep. pneumoniae	Not typeable	Alpha	Bile-soluble; inhibited by optochin
Viridans group Streptococci	Not typeable	Alpha	Not bile-soluble; not inhibited by optochin

A. *STREPTOCOCCUS PNEUMONIAE* = PNEUMOCOCCUS

1. ***Strep. pneumoniae* is a Gram-positive, lancet-shaped diplococcus.** It is **alpha-hemolytic,** with growth **inhibited by optochin** and usually **lysed by bile.**
2. **Virulence. Capsular polysaccharide** (85 serotypes) **inhibits phagocytic uptake,** unless specific antibody is present. **Pneumolysin O** (a hemolysin), **released from autolyzed *Strep. penumoniae,* damages respiratory epithelium** and stimulates outpouring of fluid, red blood cells, and leukocytes from alveoli, leading to the **productive cough.**
3. **Epidemiology.** Nasopharyngeal mucosal colonization (up to 30% of normal people) leads to a **respiratory droplet spread;** however, pneumococcus is not considered highly contagious, because disease rarely occurs in healthy people.
4. **Diseases**
   a. **Pneumococcal pneumonia** occurs as a **typical lobar pneumonia.** It is the **most common pneumonia in alcoholics and older adults** and a major cause of death. **Blood cultures are often positive.**
      i. Symptoms include **rapid onset of chills and fever and productive cough (blood-tinged, pink, or rusty sputum).**
      ii. Predisposing factors are **influenza or other respiratory viruses** that damage the respiratory mucociliary elevator, **chronic obstructive pulmonary disease (COPD), and alcoholism.**
   b. **Meningitis. *Strep. pneumoniae*** is the **major cause of meningitis in adults.** Peptidoglycan/teichoic acid elicits a strong inflammatory response in the central nervous system, resulting in elevated cell counts in cerebrospinal fluid (CSF).
   c. **Otitis media. *Strep. pneumoniae* is the dominant cause of otitis media.**
   d. **Bacteremia occurs in pneumonia and precedes meningitis.**
5. **Lab identification. Pneumococcus is an alpha-hemolytic diplococcus that is inhibited by optochin and lysed by bile.** It is typed using capsular antibodies **(Quellung reaction). Gram stain of CSF sediment** and **latex particle agglutination (LPA) tests for CSF capsular antigen** are rapid diagnostic tests for meningitis.
6. **Drug resistance of pneumococci to penicillins** is increasing through **mutations in PBPs** (transpeptidases and carboxypeptidases).
7. **Vaccination.** For people who are asplenic or >65 years old, reduction in pneumococcal infection is achieved with the pneumococcal vaccine, which has **23 different capsular polysaccharide serotypes.** For infants, the **new heptavalent**

pneumococcal conjugate vaccine (seven capsular polysaccharides linked to protein) reduces bacteremia, meningitis, and otitis media.

B. *STREPTOCOCCUS PYOGENES* = GROUP A STREPTOCOCCI = GAS

1. ***Strep. pyogenes*** is a **Gram-positive, catalase-negative coccus occurring in chains; it is beta-hemolytic, bacitracin-sensitive,** and PYR test positive.
2. **GAS virulence factors** include the **hyaluronic acid capsule (antiphagocytic but nonimmunogenic polysaccharide), M proteins (immunogenic adhesins** on the lipoteichoic acid fimbria), **protein F (binds to fibronectin), streptolysin O (immunogenic hemolysin** that stimulates **a**nti-**s**treptolysin **O** [ASO] titer), **C5a peptidase** (interferes with PMN attraction), and **hyaluronidase** and **kinases** (both involved in group A strep's ability to spread).
3. Reservoir is the **oropharynx of human carriers** (people with partial immunity).
4. **Diseases**
   a. **"Strep throat."** Symptoms are **pharyngitis with tonsillar exudate, anterior cervical lymphadenopathy, fever, and sometimes nausea.**
   b. **Scarlatina or scarlet fever.** "Strep throat" with rash is called *scarlatina* or, if severe, *scarlet fever.* ***Strep. pyogenes*** **erythrogenic toxins A and C (SPE-A and SPE-C) are pyrogenic superantigens that cause fever, rash, T cell proliferation, and B cell suppression. SPE-A and SPE-C toxins are phage-coded.**
   c. **Streptococcal impetigo** is usually characterized by **honey-crusted skin lesions** (versus *S. aureus* impetigo with bullae) and may lead to **acute glomerulonephritis if it is a certain serotype of M protein (commonly M12).**
   d. **Necrotizing fasciitis.** The **kinases, hyaluronidase, cytolytic enzymes, SPE-A and SPE-C,** and **C5a peptidase** of *Strep. pyogenes* cause a rapid, life-threatening infection.
   e. Other *Strep. pyogenes* infections include **erysipelas** (infection of dermal lymphatics, usually facial), **osteomyelitis, toxic shock–like syndrome, and puerperal sepsis.**
5. **Poststreptococcal sequelae**
   a. **Acute glomerulonephritis (dark urine, hypertension, and edema)** is usually an M12 serotype and a sequela to either pharyngitis or impetigo.
   b. **Rheumatic fever (fever, carditis, subcutaneous nodules, polyarthritis, and chorea)** may follow untreated GAS **pharyngitis** (not impetigo) and is triggered by cross-reaction between *Strep. pyogenes* M proteins and heart and joint antigens.
6. **Lab identification of *Strep. pyogenes* disease**
   a. For **pharyngitis, rapid antigen test; if negative, culture.** Carriers have negative ASO titers with repeated positive throat cultures.
   b. For invasive disease, culture on blood agar. GAS is inhibited by bacitracin.
   c. **For rheumatic fever, ASO titers >200 are positive.**
7. **Prevention.** Beta-lactams are used as prophylaxis against recurring infections in persons with rheumatic fever who are frequently exposed to children.

C. *STREPTOCOCCUS AGALACTIAE* = GROUP B STREPTOCOCCI = GBS

1. ***Strep. agalactiae* is Gram-positive, β-hemolytic,** and **bacitracin-resistant.** The **polysaccharide capsule** has terminal **sialic acid** groups, making it a poor immunogen; it binds serum protein factor H, reducing opsonization by the alternative complement pathway and decreasing phagocytic uptake.
2. **Epidemiology. GBS colonizes genitourinary (GU) and gastrointestinal (GI) tracts** and may be **sexually transmitted.**

3. **Disease**
   a. **GBS in adults:** Asymptomatic carrier, mild febrile illness, or UTI, which, in pregnant women, may lead to amnionitis or endometritis.
   b. **In neonates, GBS** has two presentations: **early onset (1–7 days),** characterized by **respiratory problems, sepsis, pneumonia, and meningitis;** and **late onset,** characterized by **septicemia and meningitis.** Intrapartum penicillin for women who test GBS+, had previous GBS+ baby, intrapartum fever of unknown origin, or who deliver at <37 weeks or >18 hours after membrane rupture reduces the incidence of neonatal disease.

D. VIRIDANS STREPTOCOCCI (*STREP. SALIVARIUS, STREP. MUTANS, STREP. SANGUIS,* ETC.)

1. **Viridans streptococci** are **α-hemolytic** (partial/"green") and **insensitive to bile and optochin.** They are **normal oral, GI, and GU tract flora.**
2. **Diseases. Endocarditis (usually subacute)** may follow **dental trauma** in persons with **previously damaged valves** not treated prophylactically. ***Strep. mutans*** causes **dental plaque and decay** through production of dextran biofilm and of acids that damage dental enamel.

III *Enterococcus* (formerly *Streptococcus*)

***Enterococcus faecalis* and *Enterococcus faecium* are normal GI tract/vaginal flora** that cross-react with **Lancefield group D** antisera. They rarely cause infections in healthy people but are **major nosocomial agents** causing disease, including **septicemias, wound infections, UTIs,** and **subacute endocarditis (in patients with previously damaged hearts following urinary or GI tract manipulations),** in compromised patients. There is increasing drug resistance. **Vancomycin-resistance** is due to modification of cell wall pentapeptide precursor from **–d-ala–d-ala to –d-lactate–d-lactate.** The latter continues to be cross-linked in the presence of vancomycin.

IV *Peptostreptococcus*

Peptostreptococci are obligate anaerobic cocci in chains found in the colon. They play a role in peritonitis.

Chapter 9

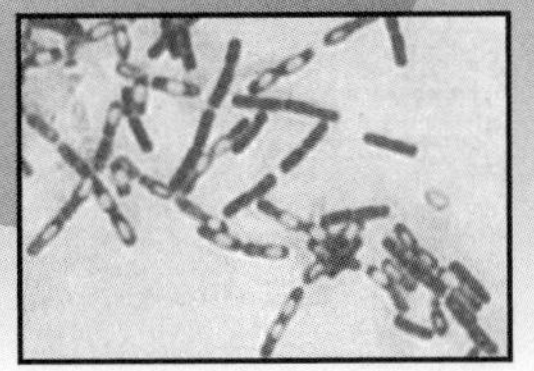
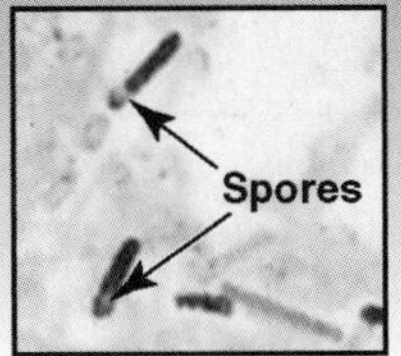

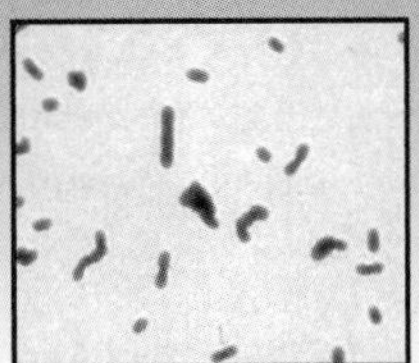

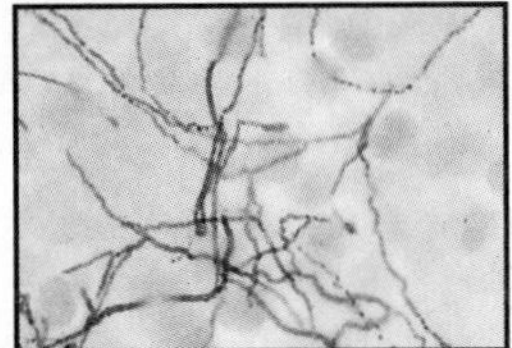
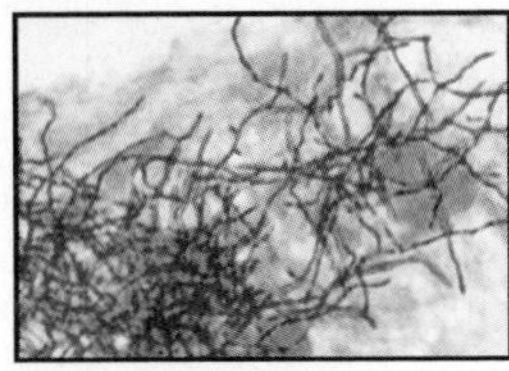
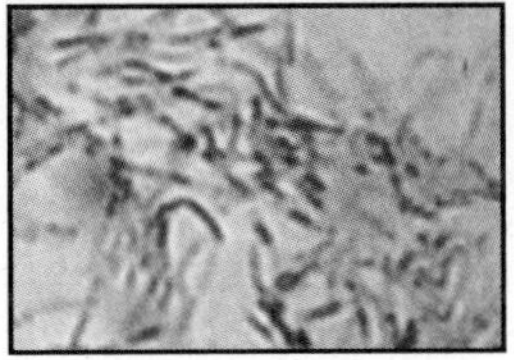

Gram-Positive Bacilli

Bacillus, Clostridium, Listeria, Corynebacterium, Actinomyces, Nocardia, Mycobacterium

I Major Genera

The major genera of Gram-positive bacilli are listed with some of their distinguishing characteristics in Table 9-1 and are described below.

II *Bacillus*

Bacillus consists of **Gram-positive, spore-formers** that all grow in ambient air.

A. *BACILLUS ANTHRACIS*

1. **Epidemiology. Spores may survive decades in soil and on animal skins** and may be traumatically implanted or inhaled. The United States has reduced animal infection by vaccination.
2. **Virulence**
   a. **Polypeptide (poly-D-glutamate) capsule (immunogenic and antiphagocytic)**
   b. **Anthrax toxin, a three-component exotoxin:**
      i. **Protective antigen (PA) (equivalent to a B component) binds to cells and facilitates the entry** of either or both lethal factor or edema factor.
      ii. **Lethal factor (LF) is protease of MAP kinase (Mitogen-activated protein kinase)**
      iii. **Edema factor (EF) is a calmodulin-activated adenylate cyclase.**

TABLE 9-1 DIFFERENTIATION OF GRAM-POSITIVE BACILLI AND BRANCHING BACTERIA

Genus	Spores?	Oxygen Utilization*	Acid-fast?	Other Features
Bacillus	Yes	Aerobe	No	
Clostridium	Yes	Anaerobe	No	
Listeria	No	Aerobe	No	Intracellular; tumbling motility
Corynebacterium	No	Aerobe	No	
Actinomyces	No	Anaerobe	No	
Nocardia	No	Aerobe	Partial; yes	
Mycobacterium	No	Aerobe	Yes	Poorly Gram staining; Often intracellular in tissue

*There are species variations with respect to growth conditions. For the USMLE Step 1, work with the large picture only.

3. **Infections**
   a. **Cutaneous anthrax** (hazard for people working with hooved animals, their skins, or wool). **Traumatic implantation of spores leads to a raised red tumorlike lesion, which progresses to a black, necrotic lesion** (eschar) with a red, rolled edge. Systemic symptoms may be present.
   b. **Anthrax pneumonia** (wool sorter's disease) is a rapid, **life-threatening hemorrhagic lymphadenitis and pneumonia** if not treated promptly.

B. *BACILLUS CEREUS* occurs naturally in rice and vegetables. The spores are not killed by boiling; when rice is made into a higher-protein food (typically **fried rice**) and is then poorly held and refrigerated, **emetic enterotoxins** are produced. **When ingested, the toxins cause rapid (1–6 hours) nausea, vomiting, and diarrhea,** similar to symptoms from staphylococcal food poisoning.

III *Clostridium*

Clostridium is a genus of **large, Gram-positive, anaerobic spore–forming rods.**

A. *CLOSTRIDIUM TETANI*

1. **Epidemiology.** ***C. tetani*** is a **soil spore former. Traumatic implantation into tissues with low oxygenation** (e.g., via puncture wounds, burn wounds, unsterile surgery, or deliveries) **leads to spore germination, growth, and toxin production.**
2. **Infection and toxicity. Tetanus toxin, a neurotoxic exotoxin, acts on anterior horn cells, blocking release of inhibitory mediators (glycine and GABA) and resulting in rigid spasm.** Spastic paralysis begins in the jaw area **(trismus, lockjaw, risus sardonicus)** and descends, if untreated, causing paralysis of large muscle groups, including opisthotonos (rigid back spasm) and death from paralysis of throat and respiratory muscles.
   a. **Treatment of symptomatic tetanus** includes
      i. **Tetanus immune globulin (TIG)** at the wound site
      ii. Vaccination at a distant site with DTP or DTaP (toxoids + acellular pertussis components) if <7 years, with new Tdap if 11–15 years but no Td yet, or Td (absorbed tetanus-diptheria toxoids) for >15 years.

TABLE 9-2 PROPHYLACTIC TREATMENT AGAINST TETANUS

Patient's Vaccination Status	Minor, New Wounds with Healthy Surrounding Tissue	More Serious Wounds
Unknown or not completed primary series of three	Vaccine	TIG around site and vaccine
Completed primary vaccination series	No vaccine; however, if >10 years since last vaccination, give booster	Booster if >5 years since last inoculation

KEY: TIG, human tetanus immunoglobin.

iii. **Spasmolytic drugs**
iv. **Metronidazole or penicillin**

b. **Prevention**
   i. The tetanus **vaccine (toxoid, the inactivated toxin)** is given at 2, 4, 6, and 18 months; at 4–6 and 11–12 years; and then as a **booster every 10 years.**
   ii. **Wound prophylaxis involves proper wound care, plus the treatment listed in Table 9-2.**

B. *CLOSTRIDIUM PERFRINGENS*

1. **Epidemiology.** ***C. perfringens*** is a large **spore former found in soil, dust, and feces.** It causes **wound infections and food poisoning (via reheated meat dishes).**
2. **Myonecrosis (gas gangrene)** is mixed anaerobic cellulitis in tissues with compromised oxygen supply. The numerous toxins of ***C. perfringens*** cause **pain, massive tissue destruction with production of gas, and shock. Tissue destruction is largely a result of alpha-toxin (a.k.a. phospholipase C), a lecithinase that destroys eukaryotic membranes.** Theta-toxin destroys polymorphonuclear neutrophil leukocytes (PMNs). Fermentation of muscle carbohydrates produces gases. If not treated, the infection is rapidly fatal.
   a. **Diagnosis** is made by visualizing **Gram-positive bacilli** in tissues with sparse PMNs; infection is **confirmed by anaerobic cultures.**
   b. Treatment is by debridement, use of hyperbaric chamber, management of shock, plus penicillin and clindamycin.
3. ***C. perfringens*** **food poisoning presents with secretory diarrhea** with acute midepigastric cramping **8 to 24 hours after ingestion of heavily contaminated meats (usually prepared in large quantities and kept warm,** not hot, for an extended time). It resolves within 24 hours.

C. *CLOSTRIDIUM BOTULINUM.* This **spore former, found in soil and dust,** causes botulism; the vegetative cells produce the neurotoxin.

1. **Food-borne botulism (toxicosis)**
   a. **Epidemiology.** The **toxin is produced in poorly canned alkaline vegetables** (e.g., green beans, mushrooms). If the contaminated food is not heated thoroughly to 60°C (140°F) to inactivate the toxin (or not heated at all, as in five-bean salad), the ingested **botulinum toxin is absorbed in the gut and acts on the myoneural junctions, blocking release of acetylcholine.**
   b. **Clinical manifestations. Incubation is 1 to 2 days,** followed by such **early symptoms as double vision, diplopia, and dry mouth, and then by symmetric, descending paralysis.**
   c. Diagnosis is made by **demonstration of toxin in remaining food,** serum, or stool and by culture of stool or food.
   d. **Treatment.** Patients require supportive care, plus hyperimmune human globulin.

2. **Infant botulism (a toxi-infection)**
   a. **Epidemiology. Infants ingest spores (found in dust or honey).** The immature gastrointestinal (GI) tract flora permit germination of the *C. botulinum* spores and growth of vegetative cells. **Toxin is produced in the GI tract.**
   b. **Symptoms** include **constipation and generalized weakness, with weak crying, poor feeding, lethargy, and loss of head control (floppy baby syndrome); it can lead to respiratory arrest.**
   c. **Treatment. Monitoring with supportive care leads to complete recovery. Hyperimmune human immunoglobulin is used.** Antibiotics are usually not used, as they may prolong the infection by preventing development of normal flora.
3. **Wound botulism (toxi-infection of a sterile site)** is infection of tissue with production of botulinum toxin; treatment is with antibiotics and hyperimmune serum.

D. *CLOSTRIDIUM DIFFICILE*

1. **Reservoir.** This **anaerobic spore former is found in soil and the human GI tract. *C. difficile* overgrowth of the GI tract is often associated with antibiotic use.**
2. Toxins are polypeptides **A (an enterotoxin)** and **B (a cytotoxin causing cytoskeleton changes in tissue culture cells, triggering a PMN response).**
3. Symptoms of infection range from **mild diarrhea to pseudomembranous colitis** (abdominal cramping, fever, and diarrhea containing blood and pus).
4. Treatment is with metronidazole not vancomycin (to avoid vancomycin resistance).

IV *Listeria*

Listeria are **Gram-positive rods with tumbling motility and cold growth** (rare in pathogens). ***Listeria monocytogenes,*** the important pathogenic species in humans, **is a facultative intracellular species.** It grows in nonimmune macrophages and cells lining the GI tract.

A. EPIDEMIOLOGY. ***L. monocytogenes*** **is found in the GI tract of animals and survives in soil. It is food-borne in cabbage, deli meats, and some soft cheeses.**

B. VIRULENCE. ***L. monocytogenes*** **is invasive; listeriolysin permits *Listeria* to escape the phagosome before lysosome-phagosome fusion occurs. *Listeria* moves laterally cell to cell by polymerization of actin filaments,** propelling the *Listeria* into the adjoining cell.

C. INFECTIONS

1. **Mild gastroenteritis** that usually occurs in the summer; some people remain fecal carriers.
2. **Septicemia in pregnant women.** *Listeria* can cross the placenta to cause **granulomatosis infantiseptica**; it can also contaminate the birth canal and cause **neonatal septicemia or, rarely, neonatal meningitis.**
3. **Meningitis in immunocompromised patients,** particularly renal transplant patients, is the most common clinical listeriosis.

V *Corynebacterium*

Corynebacterium is a genus of **Gram-positive, club-shaped, nonmotile rods that grow best in oxygen. *C. dipthteriae*** is the major pathogen.

A. **EPIDEMIOLOGY. *C. dipththeriae* has human reservoirs, with respiratory spread.** It colonizes but does not invade the oropharynx.

B. **VIRULENCE** is due to diphtheria toxin production and toxin circulation. **Diphtheria toxin (an A-B toxin) is an ADP ribosyl transferase that binds to eukaryotic elongation factor 2, inhibiting protein synthesis. Diphtheria toxin's B (binding) component "directs" the toxin primarily to the oropharyngeal mucosa, heart, and nerve cells. Toxin is only produced by *C. dipththeriae* infected with corynephage beta.**

C. **SYMPTOMS OF INFECTION. Diphtheria symptoms are pharyngitis with dirty white pseudomembrane (dead cells, fibrin, and gray pigment), "bull neck" (cervical lymphadenitis), myocarditis, cardiac dysfunction, and laryngeal nerve palsy.** Death may occur from respiratory obstruction or cardiac failure.

D. **PREVENTION OF INFECTION** is accomplished by **vaccination with diphtheria toxoid.**

E. **TREATMENT requires both antitoxin and antibiotic** (erythromycin or penicillin).

VI *Actinomyces*

***Actinomyces* are Gram-positive, non-acid-fast, anaerobic bacteria that vary from rods to branching filamentous forms.**

A. **RESERVOIR. *Actinomyces* are normal bacterial mucosal flora found in gingival crevices and the female genital tract.**

B. **INFECTIONS. Endogenous infections arise from trauma that compromises tissue blood flow and allows *Actinomyces* to penetrate. *Actinomyces israeli*** (the major pathogen) grows without respect for internal anatomic barriers. Disease is usually **cervicofacial ("lumpy jaw" from tooth extraction),** thoracic, or abdominal (e.g., from intrauterine devices). Symptoms include **abscesses, swelling, and, ultimately, sinus tract formation with hard, yellowish microcolonies called *sulfur granules*.** Brain abscesses can also occur.

C. **TREATMENT** with an antibacterial agent (usually penicillin) is usually slow but successful.

VII *Nocardia*

***Nocardia* are Gram-positive, aerobic, partially acid-fast filaments that break up into rods.** These hardy organisms are found in **soil.**

A. **PULMONARY NOCARDIOSIS (acquired by inhalation) occurs in patients with low white blood cell or CD4+ counts.** Infections resemble tuberculosis with hematogenous spread to other organs, including brain. **Disease is not contagious.**

B. **CUTANEOUS OR SUBCUTANEOUS INFECTIONS** may be caused by traumatic implantation.

C. **BRAIN ABSCESSES** can be caused by ***Nocardia* or *Actinomyces*.** (See Table 9-1 to distinguish these two organisms.)

VIII Mycobacteria

Mycobacteria are **poorly Gram staining, obligate aerobic bacilli. Mycobacteria are acid-fast (because of the very waxy and hydrophobic arabinogalactan-mycolate cell wall layer) and are usually intracellular.**

A. *MYCOBACTERIUM TUBERCULOSIS* is a **facultative intracellular, human pathogen that causes tuberculosis** (TB). *M. tuberculosis* is a major problem among the **poor and HIV-positive who live in crowded urban living situations,** because of increased chance of respiratory spread and presence of untreated patients.

1. Virulence is not clearly understood, but compounds in the waxy *M. tuberculosis* cell envelope play a role. Cord factor (trehalose mycolate) inhibits mitochondrial respiration (and, in culture, causes virulent *M. tuberculosis* to grow as serpentine cords). Sulfolipids (a.k.a. sulfatides) inhibit phagosome-lysosome fusion, allowing *M. tuberculosis* to survive intracellularly. *M. tuberculosis* cell wall lipoarabinomannan is functionally analogous to lipopolysaccharide. Pathogenesis depends on the balance of two related host responses: the *M. tuberculosis* killing of nonactivated macrophages and cell-mediated immune activation of macrophages to destroy ingested bacteria. Granuloma formation often walls off still viable *M. tuberculosis.*
2. **Exposure, infection, and disease (TB).**
   a. **Primary TB.** Typically, a healthy person who is infected with *M. tuberculosis* has limited replication of the organism in a lung "spot" and adjoining lymph node, both of which usually become "healed" off in granulomas (the Ghon complex) without active disease symptoms. The organism's growth is slowed down by the reduced oxygen level in the granuloma, but *M. tuberculosis* may remain viable without isoniazid (INH) treatment.
   b. **Reactivational disease.** Without treatment, if the individual's cell-mediated immune system declines, a granuloma's wall may erode (freeing the organisms into higher levels of oxygen); reactivational (secondary) TB occurs. If a person's immune system is initially unable to contain the infection, contiguous spread (sometimes with cavitary disease) may result. Hematogenous dissemination results in systemic disease (miliary TB).
3. **Diagnostic tools**
   a. **Tuberculin skin test.** Purified protein derivative (5TU PPD) is injected intradermally. Induration is read at 48 to 72 hours. A level of ≥10 mm is positive in a person with risk of exposure; ≥5 mm is positive in HIV-positive patients. However, a positive PPD test only indicates exposure; it cannot, by itself, distinguish exposure from active disease. In addition, in patients with overwhelming disease, the test may be negative.
   b. **Chest auscultation and x-ray,** along with clinical symptoms, aid diagnosis.
   c. **Sputum microscopy.** Rhodamine-auramine stain of clinical specimens (e.g., sputum) is used. This sensitive fluorescent dye binds to the waxy cell wall of the mycobacteria, but it is not specific, because no antibodies are involved. If rhodamine-auramine stain is positive, confirm with acid-fast stain, culture the sputum for mycobacteria, and run drug susceptibilities.
   d. **Culture.** Acid-fast bacilli (or auramine-rhodamine fluorescent stain confirmed by acid-fast stain) in concentrated sputum (but not urine) suggests mycobacterial disease. Standard media used to culture *M. tuberculosis* include Lowenstein-Jensen, Middlebrook, and palmitic acid-containing broths for rapid automated systems. *M. tuberculosis* produces niacin (most other mycobacteria do not). It has a heat-sensitive catalase, so in the standard

catalase test run at 68°C (154°F), it is catalase-negative. Antibiotic-susceptibility testing is important.

4. Treatment for active TB always involves use of multiple drugs. (*M. tuberculosis* develops drug resistance rapidly when treated with a single drug.)
   a. **Uncomplicated pulmonary tuberculosis.** Current standard protocol is to start with three drugs (INH, rifampin [RIF], and pyrazinamide) for 2 months, and then to cut back to INH and RIF for 4 more months or until sputum smear and culture are negative for 2 consecutive months.
   b. **Reasonable risk of infection with multiple-drug-resistant TB.** Ethambutol or streptomycin is added to the above regimen for the first 2 months.
5. Treatment for recent tuberculin skin test converters (sign of primary asymptomatic TB) <35 years of age is INH for 6 to 9 months.

B. *MYCOBACTERIUM AVIUM-INTRACELLULARE* is **an opportunistic soil or water organism** that causes infection in compromised hosts. **Pulmonary infections, which occur in patients with cancer, organ transplant, and AIDS,** are similar to TB. ***M. avium-intracellulare* is not contagious from person to person.** Prophylaxis is routine in AIDS patients when CD4+ cells are <50 per mm^3.

C. *MYCOBACTERIUM MARINUM* is a marine aerobe that causes **fish tank granulomas.**

D. *MYCOBACTERIUM LEPRAE,* an **obligate intracellular pathogen** (no culture), invades skin, peripheral nerves, and, in lepromatous leprosy, upper airways and nasal mucosa. **Humans are the only significant reservoir (there are some infected armadillos).** Leprosy is a disease with a spectrum of symptoms. **Features of the two extreme forms are presented in Table 9-3.**

TABLE 9-3 TUBERCULOID VS. LEPROMATOUS LEPROSY

Distinguishing Features	Tuberculoid Leprosy	Lepromatous Leprosy
Cell-mediated immunity (CMI)	Strong CMI	Weak CMI
Lepromin skin test	Lepromin-positive	Lepromin-negative
Tissue *M. leprae* (as seen in acid-fast stained punch biopsy)	Few AFB seen	Many AFB seen (foam cells filled) More contagious
Symptoms	• One to few flat lesions • Nerve enlargement • Loss of sensation leads to burns, trauma	• More severe disease • Multiple, bilaterally distributed skin lesions (often nodular) • Leonine facies

KEY: AFB, acid-fast bacilli.

Chapter 10

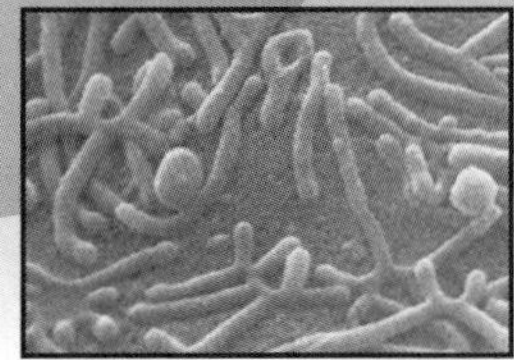

Mycoplasmas

Mycoplasma, Ureaplasma

I Mycoplasmas

Mycoplasmas lack peptidoglycan. Although they can incorporate sterols into their membranes, they cannot make them; thus, they **require sterol-containing media.** Mycoplasmas are **tiny extracellular bacteria not seen on Gram stain.**

A. ***MYCOPLASMA PNEUMONIAE*** is an **extracellular human pathogen** with respiratory spread. It **adheres via P1 protein to** sialoglycolipids on **ciliated respiratory mucosal cells, but it does not invade.** Instead it secretes chemicals such as hydrogen peroxide, **causing ciliastasis** and ultimately **respiratory epithelial desquamation** and a **prominently lymphocytic inflammatory response.**

1. **Infections. *M. pneumoniae* causes sore throat, bronchitis, and/or otitis media** and **atypical bronchopneumonia** (a.k.a. "walking pneumonia"), all without coryza. Infection starts with the upper respiratory tract and spreads to the lower. It begins with headache, fever, and malaise and ultimately causes bronchopneumonia, characterized by a hacking cough (initially dry, later productive), which lasts up to 4 weeks. ***M. pneumoniae*** is the **leading cause of pneumonia from 5 years to 15 years** but can infect individuals of any age. (*Viral pneumonia predominates* in infants, and *Streptococcus pneumoniae* in adults >65 years.)
2. Diagnosis was often clinical and serological because culture is slow, but now polymerase chain reaction is used. Complement fixation or enzyme-linked immunosorbent assay (ELISA) titers ≥ 1:32 strongly suggest infection. About 65% of patients produce cryoagglutinins (which agglutinate red blood cells [RBCs] in the cold) due to cross-reaction of P1 with I antigen on RBCs. Test is suggestive but not specific.
3. Treatment is with macrolide or tetracycline, not beta-lactam antibiotics.

B. ***MYCOPLASMA HOMINIS*** is a genitouropathogen causing postpartum or portabortal fever.

C. ***UREAPLASMA UREALYTICUM*** is also missing a cell wall. It **metabolizes urea** and causes **nongonococcal, nonchlamydial urethritis and prostatitis,** both of which often present with fever.

II Other Bacteria Not Reliably Seen on Gram Stain

A. MYCOBACTERIA do not take up the stain well because of their waxy cell wall, but they are basically Gram-positive (no outer membrane or lipopolysaccharide).

B. SEVERAL ORGANISMS (e.g., **spirochetes, rickettsias, and chlamydiae**) are so small that the trapped stain color is not reliably seen.

C. *LEGIONELLA* stains Gram-negative only if the counterstain time is increased.

Chapter **11**

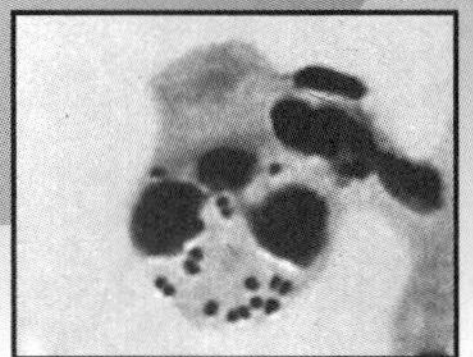

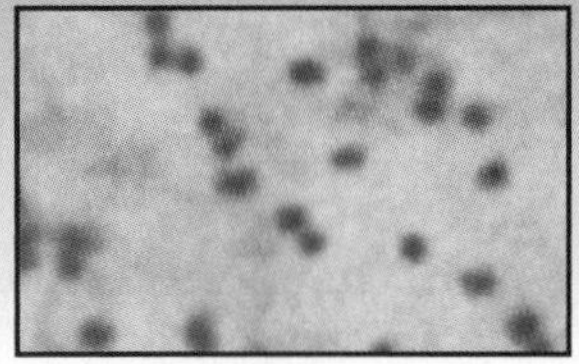

Gram-Negative Cocci

Neisseria, Moraxella

I *Neisseria*

Neisseria are **Gram-negative**, pyogenic, **kidney bean-shaped diplococci** that are **oxidase-positive.** In the presence of cytochrome C oxidase, the reagent (phenylenediamine) quickly turns black. The outer membrane contains **a modified lipopolysaccharide (LPS)** called **lipo-oligosaccharide (LOS)**, which has endotoxic activity and is antigenic. Along with several commensal *Neisseria,* there are two major pathogens: *N. meningitidis* and *N. gonorrhoeae.* Both pathogens bind to and invade epithelial cells, sequester iron by transferring binding to surface receptors, and have serum resistance.

A. ***NEISSERIA MENINGITIDIS*** (a.k.a. **meningococcus**) has a **capsule** and **ferments maltose.**

1. **Epidemiology.** The only reservoir is the **human nasopharynx;** transmission is by respiratory spread. (About 10% of the healthy carry *N. meningitidis* in their oropharynx.) Peak incidence of disease is 6 months to 2 years, with a second outbreak in young adults associated with college bars and residence halls.
2. **Virulence.** Virulent *N. meningitidis* must colonize and invade the mucosa, then enter the bloodstream, surviving to ultimately cross the blood–brain barrier to cause meningitis.
   a. **IgA protease** aids in colonization. **Pili are involved in mucosal adherence and invasion.** Virus or mycoplasma upper respiratory tract (URT) infection may increase risk.
   b. **Polysaccharide capsule, which reduces phagocytic uptake and complement activation, is critical to surviving in the bloodstream and reaching the blood–brain barrier.** *N. meningitidis* B and C are the most common serotypes in the United States. Because the B capsule is made of sialic acid, it has increased resistance to serum killing and is not a good immunogen. Therefore, it is not in the current polysaccharide vaccine.
   c. In vitro, ***N. meningitidis*** **produces excess outer membrane (more than can be incorporated), which is probably why petechiae and purpura are present so early in infection** and why meningococcus is such a potent inducer of disseminated intravascular coagulation (DIC) Figure 11-1.

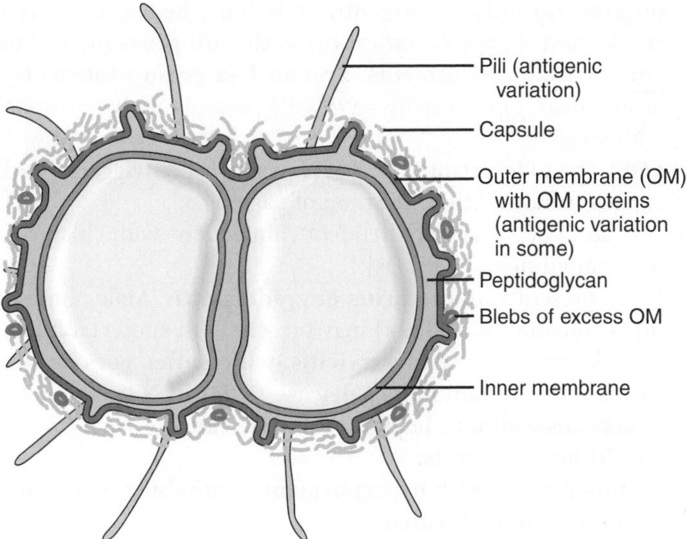

● **Figure 11-1** *Neisseria meningitidis* has both capsule and excess outer membrane fragments.

3. **Infection. Respiratory symptoms (pharyngitis or pneumonia) may precede the septicemia (meningococcemia). Early meningitis symptoms include acute onset of headache, fever, stiff neck, vomiting, and rash, which leads to rapid decline to coma and Gram-negative shock.** Death often occurs within 24 hours of onset of symptoms.
4. **Lab ID.** (Antibiotic treatment should begin within 30 minutes of suspicion of potential *N. meningitis.* Do not wait for imaging or lab results.) **Gram stain of cerebrospinal fluid (CSF)** or the less-sensitive latex particle agglutination tests for capsular antigens may rapidly detect *N. meningitidis.* **CSF and blood are cultured on chocolate agar with increased CO_2. *N. meningitidis* ferments maltose.** (Note: Nasopharyngeal cultures of close contacts in an outbreak are plated on Thayer-Martin medium; nonpathogenic strains of *Neisseria,* which are found in the URT and genitourinary [GU] tract, do not grow on Thayer-Martin medium.)
5. **Prevention**
   a. Vaccination with the *N. meningitidis* conjugate vaccine is recommended for 11–12 year olds, college freshmen, new military recruits, or in outbreaks in the general population. The vaccine has four serotypes of capsule: Y, W-135, C, and A conjugated to protein (Ⓜ YWCA vaccine). The B capsule (a polymer of sialic acid) is not immunogenic.
   b. **Rifampin, which reduces mucosal colonization,** is also used to reduce risk of disease in close contacts.

B. *NEISSERIA GONORRHOEAE* (a.k.a. gonococcus)

1. ***N. gonorrhoeae*** is a **Gram-negative diplococcus with flattened sides. It is oxidase-positive but has no capsule and does not ferment maltose.** Human carriers are the reservoir. Transmission is via sexual contact or passage through an infected birth canal. *N. gonorrhoeae* is sensitive to cold and drying.
2. **Virulence**
   a. **IgA protease** aids in colonization.
   b. **Pili** play a major role in attachment to mucosal surfaces, triggering translocation into tissues. **Pili have constant and hypervariable regions (>1 million**

piliated variants = antigenic variation); hence, there is no immunity to reinfection. Phase variation turns the *pili* genes on and off.

c. **Outer membrane proteins. Protein I (a porin protein) is associated with attachment.** Opa (opacity = OMP II) proteins are hypervariable and associated with virulence.

d. **LOS.** The LPS variant (LOS) of *N. gonorrhoeae* **triggers cytokine release and causes inflammation.** (See top of page 64.)

3. **Gonorrhea** (Note: There are frequent coinfections with chlamydiae or other sexually transmitted diseases [STDs].)
   a. Males present with **urethritis or epididymitis.** Male homosexuals (or anyone practicing oral or anal sex) may present with anorectal lesions or pharyngitis.
   b. **Females may have endocervicitis, which often goes undiagnosed and may cause pelvic inflammatory disease (PID). If N. *gonorrhoeae* is not treated, it may disseminate hematogenously, causing rash and arthritis in large weight-bearing joints.**
   c. **Infants present with hyperpurulent ophthalmia, which leads to rapid loss of eyesight if not treated.**
4. **Lab ID.** Tests include **DNA probes** and polymerase chain reaction, microscopy **looking for intracellular Gram-negative organisms in neutrophils (useful only with male urethral exudate),** or **cultures on Thayer-Martin medium** (chocolate agar with colistin, vancomycin, and nystatin ± trimethoprim incubated with high CO_2; candle jar or Gas Pak).
5. **Drug resistance. Penicillinase-producing N. *gonorrhoeae* (PPNG). Because plasmid-encoded penicillinase TEM-1 (causing high-level resistance) is now in >20% of N.g. strains, penicillin is no longer used.**
6. **Prevention. In neonates, silver nitrate or erythromycin in the eyes prevents infection that would cause loss of eyesight.**
7. **Treatment.** Ceftriaxone is used for treatment; azithromycin or doxycycline is added if coinfection with *Chlamydia trachomatis* is not ruled out.

II *Moraxella*

A. *MORAXELLA* is a genus of **Gram-negative diplococci** found in the URT as normal flora. The most significant human pathogen is *M. catarrhalis*.

B. INFECTIONS. *M. catarrhalis* causes otitis media (third after *Streptococcus pneumoniae* and *Haemophilus influenzae*), which carries a high level of drug resistance. It may cause bronchitis in patients with chronic obstructive pulmonary disease or respiratory and central nervous system infections in immunocompromised patients.

Chapter **12**

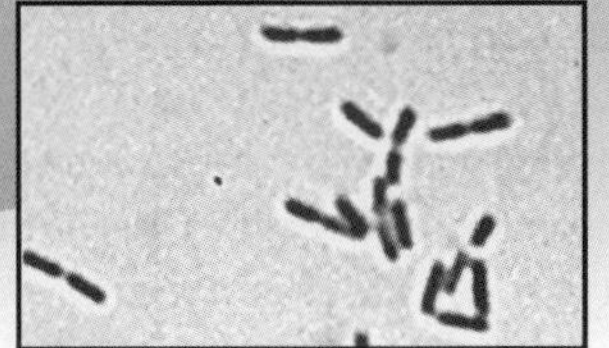

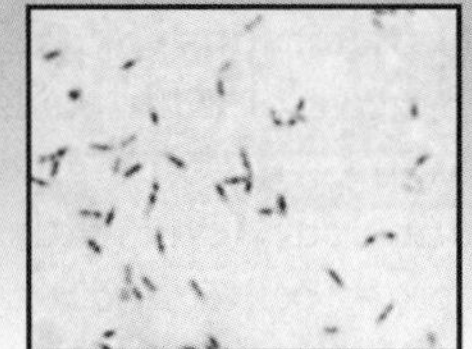

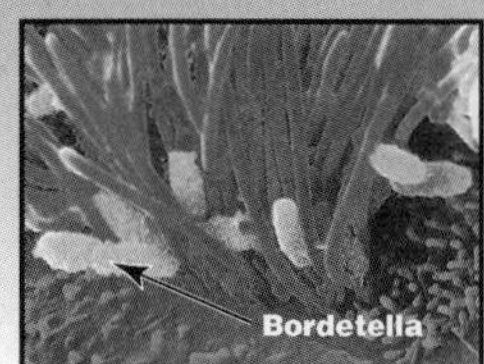

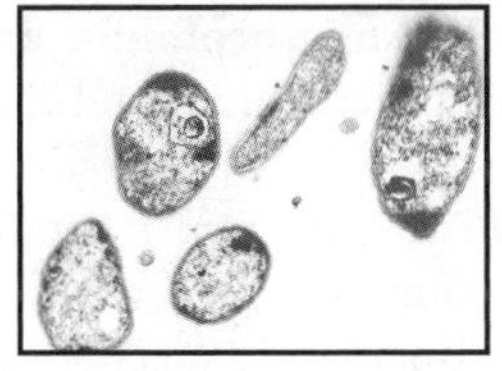

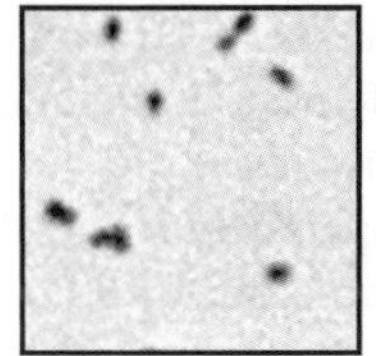

Gram-Negative Aerobic Bacilli

Pseudomonas, Legionella, Bordetella, Francisella, Brucella

I *Pseudomonas*

A. *PSEUDOMONAS* is a genus of **aerobic Gram-negative rods** that are motile and **ubiquitous in water and soil.** ***P. aeruginosa*** is a major nosocomial and **opportunistic pathogen.** *P. aeruginosa* has a distinctive **grapelike odor,** often produces pigments (the **blue-green pyocyanin** and fluorescein), and grows on most media.

B. VIRULENCE. In addition to the **pigments** and **endotoxin,** *Pseudomonas* has a **polysaccharide slime layer (alginate) that is antiphagocytic** and involved in adherence/colonization; **exotoxin A (an ADP ribosyl transferase that inactivates elongation factor 2),** which shuts down mammalian protein synthesis primarily in liver cells; and **an elastase,** which damages immunoglobulins, elastin, and some collagens.

C. RESERVOIRS. Sources of *Pseudomonas* include water faucet aerators, drains, respiratory equipment, raw vegetables, flowers, slimy bar soap, and standing water (distilled or tap).

D. INFECTIONS. Exposure to *Pseudomonas* is so common that at any point in time about **10% of the normal population has transient *Pseudomonas* gastrointestinal (GI) tract colonization** (resulting only in loose stools). This incidence rises to **70% in hospitalized patients on antibiotics.** Neutropenia, cystic fibrosis, and burns are potent predisposing conditions.

1. ***Pseudomonas* infections in normal immuno*competent*** people include **eye ulcers** from trauma or extended wear of contact lenses; **wounds** (e.g., puncture

wounds through soles of tennis shoes, a *P. aeruginosa* haven!); **swimmer's ear; and hot tub folliculitis** (colonized or inflamed hair follicles from the neck down).

2. **Burn patients.** GI tract colonization, often from raw vegetables or poor disinfection of equipment like faucets or whirlpools, leads to colonization of burns and may result in frank **cellulitis with blue-green pus** and **septicemia.** Symptoms of *P. aeruginosa* septicemia include **fever with or without ecthyma gangrenosum** (lesions with a black necrotic center and a raised erythematous margin) and **may progress to Gram-negative shock and death.**
3. **Cystic fibrosis (CF) patients. Pulmonary colonization** with *P. aeruginosa* occurs early in CF patients and is difficult to eradicate because **microcolonies are protected in the capsular slime.** Repeated pneumonias are common, and *P. aeruginosa* is often the cause of death in CF.
4. **Leukemic, transplant,** or **neutropenic patients** (<500 PMNs per mm^3). **Septicemia, pneumonia** (especially if on ventilator), or both are common.
5. Patients with long-term urinary catheters are at risk for urinary tract infections (UTIs).
6. **Intravenous drug abusers** are susceptible to endocarditis, osteomyelitis, and arthritis.
7. **Diabetics are at risk for malignant otitis externa,** which has a high fatality rate.

E. DRUG RESISTANCE. Because *Pseudomonas* is missing some high-affinity porins, it has native resistance to many antibiotics, as well as some acquired plasmid-mediated resistance.

II *Legionella*

A. *LEGIONELLAE* are **poorly Gram-staining,** but Gram-negative, **facultative-intracellular,** aerobic bacilli. *L. pneumophila* causes about 80% of Legionnaire's disease (LD).

B. RESERVOIRS. *Legionella* is an **aquatic organism** found intracellularly in amebae in streams. It contaminates air conditioner cooling tanks in hot weather, hot water heaters, and so forth. There is **no human-to-human transmission.** Clusters of cases are from common source exposure.

C. VIRULENCE. Because *Legionella* is intracellular, it is taken into alveolar macrophages. It inhibits the lysosome-phagosome fusion, grows in the phagosome, and ultimately kills the cells.

D. INFECTIONS. LD is a necrotizing, **multifocal pneumonia** accompanied by myalgia, **headache, fever, diarrhea,** and generally **nonproductive cough.** It occurs mainly in **older people debilitated** by diabetes, alcoholism, smoking, or immunocompromise. Fatality rate is high if untreated. The same legionellae also cause Pontiac fever, a short, nonfatal pneumonitis that is probably a reaction to the inhaled components rather than infection.

E. LAB ID. Clinical suspicion includes toxic pneumonia, no sputum, and negative on Gram stain of induced sputa. Order direct fluorescent antibody **(DFA)** for *Legionella* (insensitive, but specific if positive) on bronchiole alveolar fluid **(BALF)** or lung biopsy and **culture on buffered charcoal yeast extract (BCYE) medium** (a special medium for legionellae that has the **required L-cysteine and ferric iron,** along with detoxifying charcoal). If Gram-negative rods grow on BCYE but not on blood agar, do DFA on growth to make diagnosis. **Polymerase chain reaction (PCR) for *Legionella* rRNA and urine antigen tests are now available.**

F. **TREATMENT requires antibiotics** that effectively **penetrate human cells.** Many legionellae produce penicillinases and cephalosporinase. In seriously immunocompromised patients, a macrolide is commonly used with added rifampin.

G. **PREVENTION.** Hyperchlorinate, disinfect, or heat to >70°C to decontaminate water source.

III *Bordetella*

A. ***BORDETELLA*** is a genus of **small, Gram-negative, aerobic, extracellular bacilli.** *B. pertussis* is the most important (and uniquely human) pathogen.

B. **VIRULENCE. Adhesins and toxins** are the primary mediators of the disease. Their production is co- (and up-) regulated at 37°C by temperature-sensitive membrane-regulatory protein BvgS. The organism is **not invasive.** Instead, *B. pertussis* attaches to ciliated respiratory epithelium through multiple **adhesins (filamentous hemagglutination, pertussis toxin, and an adhesin called *pertactin*)** to inhibit and then kill these cells.

1. **Pertussis toxin** is an A-B component toxin. The A component, an ADP ribosyl transferase, inhibits G_i (i.e., it **inhibits the inhibitor of adenylate cyclase**). As a result, cyclic adenosine monophosphate **(cAMP) increases in target cells.** Lymphocytosis and hypoglycemia result clinically.
2. **Adenylate cyclase, produced by *B. pertussis*,** enters human cells, where the intracellular calmodulin activates it and **causes an increase in intracellular cAMP**.
3. **Tracheal cytotoxin inhibits respiratory epithelial cilia and ultimately kills the cells.**

C. **INFECTION IN IMMUNOLOGICALLY NAÏVE.** Pertussis follows a 7- to 10-day incubation.

1. **Catarrhal stage** (1 to 2 weeks): Patient is **contagious.** Symptoms include **rhinorrhea** (copious and mucoid) and **symptoms of a cold.** Fever is usually not prominent.
2. **Paroxysmal stage** (2 to 4 weeks): Marked by **repetitive short expiratory bursts through a narrowed glottis, ending in inspiratory gasp and often vomiting. The cough in unvaccinated individuals is so severe that it frequently leads to aspiration pneumonia, eye hemorrhages, hernias, frenal ulcers, and seizures,** as well as possible permanent central nervous system (CNS) damage. Lymphocytosis is common. Bordetellae disappear by the second week.
3. **Convalescent stage leads to recovery, unless permanent CNS damage has occurred.**
4. In vaccinated patients, **more than 20% of coughs that persist >2 weeks in afebrile adults are pertussis.** Coughing until vomiting is not uncommon.

D. **LAB ID.** Tests are nasopharyngeal mucus cultures or cough plates on Regan-Lowe and direct immunofluorescence on specimens. Cultures are usually negative (even though the person has pertussis) if antibiotics have already been used, if symptoms have been present for >4 weeks, or if the individual has been vaccinated. Thus, diagnosis is often clinical. There is a **DFA for nasopharyngeal smears.**

E. **PREVENTION.** All new **DTaP vaccinations** include the **pertussis filamentous hemagglutinin** and the **pertussis toxoid;** they vary in other components. Vaccine-induced immunity wanes at about 11 years; thus, there is little to no immunity in newborns, who are at extremely high risk of exposure and serious disease until vaccinated.

IV *Francisella*

A. ***FRANCISELLA*** is a genus of **facultative-intracellular, Gram-negative bacilli.** The most important human pathogen of this genus is ***F. tularensis.***

B. **RESERVOIRS.** *F. tularensis* infects **wild mammals, often without overt disease.** Human infections start with **transfer of infected animal blood,** most commonly by trauma while **skinning a rabbit** or when a **tick** from the rabbit then **bites** the human.

C. **INFECTIONS. Ulceroglandular tularemia** (from the cuts or vector bite) begins with fever, chills, malaise, and an ulcerating papule at the inoculation site. Septicemia leads to disseminated abscesses and caseating granulomas.

D. **LAB ID. Diagnosis is by serology or immunofluorescent microscopy.** Because *Francisella* is fastidious and hazardous to culture, it is cultured only in reference labs.

V *Brucella*

A. ***BRUCELLA* is a genus of Gram-negative, facultative-intracellular bacilli that localize in the cells of the reticuloendothelial system.**

B. **RESERVOIRS.** Brucellae cause **genitourinary (GU) tract infections in goats, pigs, and cattle.** *Brucella* is spread from infected animals by direct mucosal contact, traumatic skin implantation, or ingestion. Thus, it is an **occupational hazard of veterinarians, slaughterhouse workers, farmers, and anyone who ingests unpasteurized dairy products.**

C. **INFECTIONS.** Brucellosis, or undulant fever, is an **influenzalike disease that may cause temperature undulations, with drenching sweats in late afternoon or evening.** Disease varies from **mild and self-limiting (*Br. abortus* from cattle)** to **serious (*Br. melitensis* from goats). It may be chronic (most commonly with *Br. suis* from swine), may resemble chronic fatigue syndrome,** and may be associated with **depression that lasts for years.**

D. **LAB ID.** Diagnosis is made by **serology and blood cultures.** Special handling is required; notify the lab.

Chapter 13

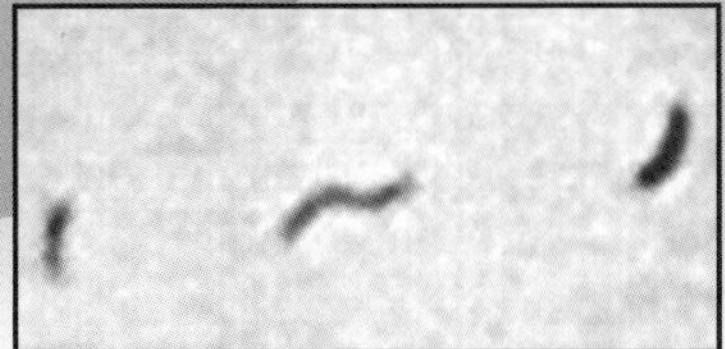

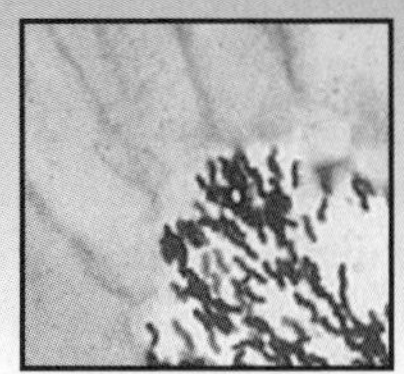

Gram-Negative Microaerophilic Curved Bacteria

Campylobacter, Helicobacter

I *Campylobacter*

A. ***CAMPYLOBACTER* are Gram-negative curved rods** with a **polar flagellum.** The organisms are often found in "nose-to-nose" pairs that look like flying seagull's wings. *Campylobacter* are **microaerophilic and fastidious** (require specific media), **grow at 42°C,** and are **oxidase-positive** nonfermenters. The most important human pathogen of this genus is *C. jejuni.* (*C. fetalis* is a problem in immunocompromised people.)

B. **EPIDEMIOLOGY.** ***C. jejuni*** is found in the **gastrointestinal (GI) tract of many wild and domestic animals,** including dogs. Human infection is most often through **ingestion of raw or undercooked poultry** or direct contact with infected pets or animals.

C. **DISEASE. *C. jejuni* is one of the most common causes of infectious diarrhea** (estimated 40%). Symptoms include **fever, acute abdominal pain** (sometimes resembling appendicitis), **malaise,** and **diarrhea, often with blood and pus.** Disease is generally self-limiting, lasting <7 days. Infection with *C. jejuni* is the leading known cause of **Guillain-Barré syndrome. Reactive arthritis** may also follow acute infections.

D. **LAB ID** is by **microaerophilic** culture on special media (Skirrow or Campy) at **42°C.**

II *Helicobacter pylori*

A. ***HELICOBACTER PYLORI* is a Gram-negative, flagellated, spiral-shaped bacterium.** It is **microaerophilic and urease-positive.**

B. **VIRULENCE. Through urease activity and the creation of an ammonia cloud around itself, *H. pylori* survives the low gastric lumen pH.** It is chemotactic to hemin found in the gastric tissue **and moves by rapid flagellar motility to the tissue where,**

with the aid of its mucinase and **twisting motility,** it traverses the gastric mucin to the tissue where it **adheres to the Lewis blood groups.** Additional virulence genes carried on a **pathogenicity island** (a section of DNA imported from another organism) code for the **Cag$^+$ protein and a type IV secretion system, which injects vacuolating toxin (VacA)** into the epithelial cells, triggering apoptosis. Inflammation is variable and has multiple causes, including the toxic effect of VacA, the ammonia, Cag$^+$ protein induction of IL-8, and the neutrophil activating protein (NAP).

C. **EPIDEMIOLOGY.** Humans only; transmission is assumed **fecal-oral, possibly during other diarrheal diseases.** Worldwide incidence by middle age is nearly universal in developing countries and 20% to 30% in developed. Worldwide, gastric carcinoma is one of our fastest growing carcinomas; rates parallel *H. Pylori* incidence.

D. **DISEASE.** ***H. pylori*** **is now known to cause gastritis, gastric and duodenal ulcers,** as well as **adenocarcinomas** and mucosa-associated lymphoid tissue **(MALT).** Some people appear to be asymptomatic carriers.

E. **LAB ID.** There are several methods of identification, including **serology for *H. pylori* antibodies, a urease breath test ($^{13}C^-$ or $^{14}C^-$ labeled urea ingested; test for "hot" CO_2),** or biopsy with microaerophilic culture at 37°C or direct urease test.

F. **TREATMENT.** To prevent relapse, **treatment must include antibiotics** (e.g., amoxicillin and clarithromycin), **omeprazole to speed healing, and sometimes bismuth.**

Chapter 14

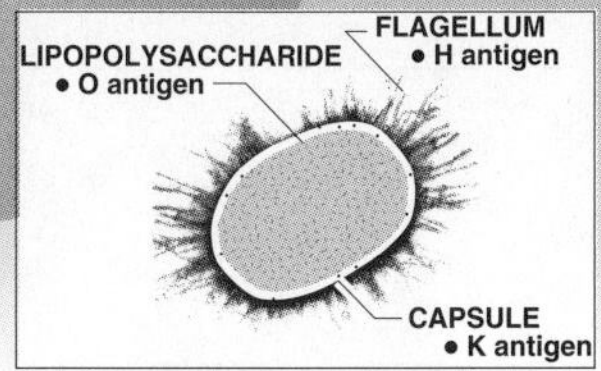

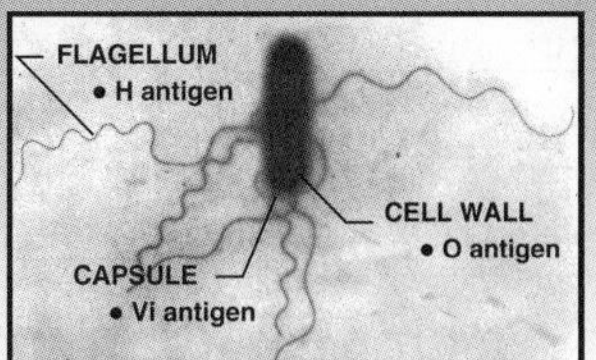

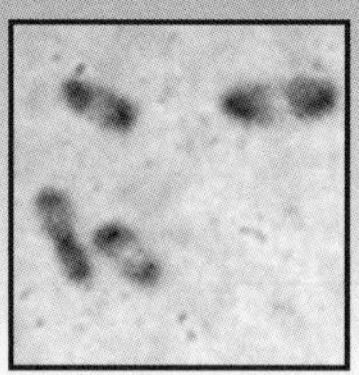

Gram-Negative Facultative Anaerobic Bacilli: Family Enterobacteriaceae

Escherichia, Klebsiella, Shigella, Salmonella, Proteus, Yersinia

I Enterobacteriaceae (Family*)

A. THE ENTEROBACTERIACEAE FAMILY consists of Gram-negative rods that are **facultative anaerobes.** All **ferment glucose, are oxidase-negative** and **catalase-positive,** and **reduce nitrates to nitrites.** High-yield genera of the Enterobacteriaceae are listed in Table 14-1.

B. VIRULENCE FACTORS. Figure 14-1 is a visual memory trick of the **three surface antigens: O (outer membrane), H (flagellar),** and **K (capsule) antigens. Pili** (which aid adherence) and **outer membrane proteins** (OMPs) may also be present. Many strains produce **exotoxins;** all produce **endotoxin.**

II *Escherichia coli*

A. *ESCHERICHIA COLI* ferments lactose. Most strains of *E. coli* are nonpathogenic, normal human intestinal flora; other strains are pathogens with differing virulence factors and effects. Enterohemorrhagic *E. coli* (EHEC) is associated with cattle feces.

B. INFECTIONS, SYMPTOMS, VIRULENCE FACTORS

1. **Cystitis** (urinary frequency with dysuria). The type 1 pili found on most *E. coli* adhere to the perineal area; *E. coli* may be mechanically (sex, catheters) aided in entry into the urethra and bladder.

*It is easier to learn the Enterobacteriaceae family characteristics and which organisms are Enterobacteriaceae than to learn characteristics, such as nitrate reduction, for each genus.

TABLE 14-1 HIGH-YIELD GENERA OF THE ENTEROBACTERIACEAE

Lactose Fermenters = CEEK	
C = *Citrobacter*	
E = *Enterobacter*	
E = *Escherichia*	
K = *Klebsiella*	
Lactose Nonfermenters = ShYPS	
Sh = *Shigella*	Nonmotile; no H_2S produced
Y = *Yersinia*	
P = *Proteus*	Highly motile; H_2S produced
S = *Salmonella*	

2. **Pyelonephritis. P pili** of **uropathogenic *E. coli*** (UPEC) strains allow adherence to the uroepithelium, allowing retention in the bladder (despite urine's flushing action) and entry to the ureters and kidneys. In the ureters and kidneys, other virulence factors (**endotoxin** and an **alpha hemolysin**) cause damage, as well as the additional symptoms of **fever, hematuria,** and **flank pain,** and create a high risk of **Gram-negative septicemia and shock.**

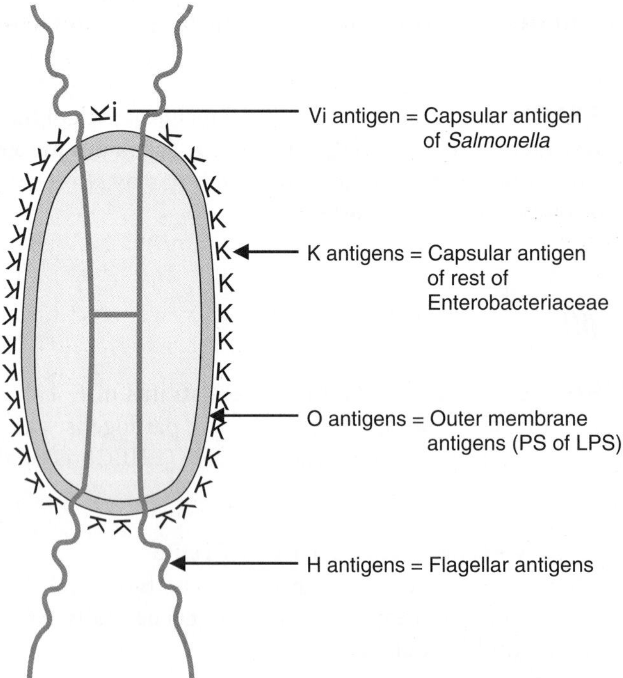

● **Figure 14-1** Enterobacteriaceae antigen memory trick (not a diagram of the structure). O antigens are the polysaccharide part of lipopolysaccharide in the bacterial cell envelope (the big "O" in the figure). H antigens are the flagella present on some Enterobacteriaceae. (You can visualize a huge letter H, with helical ends as the flagella.) K antigens are polysaccharide capsular antigens present on some Enterobacteriaceae.

a. **Lab ID.** Diagnostic methods for cystitis and pyelonephritis include
   i. **Dipstick tests,** which may show **leukocyte esterase** (sign of pus in urine; not always associated with bacteriuria), **hematuria,** or positive **nitrites**
   ii. Presence of Gram-negative bacteria on **unspun urine**
   iii. **Quantitative cultures** and susceptibilities—More than 1,000 bacteria per mL urine is considered positive in symptomatic individuals.

b. **Antimicrobial susceptibilities in pyelonephritis are important.** Empiric treatment of uncomplicated cystitis depends on local susceptibilities.

3. **Diarrheas**

a. **Enterotoxic *E. coli* (ETEC) is a major cause of "traveler's diarrhea"** and infant diarrhea in developing countries; both are watery diarrhea. ETEC is transmitted in developing countries through poor sanitation and use of **human** feces as fertilizer on food crops. Two toxins are produced in the small intestines. **The labile toxin (LT) is an ADP ribosyl transferase,** which ADP-ribosylates and **stimulates G_s** (the positive regulator of adenylate synthetase), **causing an increase in cyclic adenosine monophosphate (cAMP). This leads to loss of fluids and electrolytes, causing a watery, noninflammatory diarrhea.** The stable toxin (ST) stimulates cyclic guanosine monophosphate (cGMP), causing a similar reaction. Treatment is **fluid and electrolyte replacement** therapy.

b. **Enteropathogenic *E. coli* (EPEC)** is a major cause of **chronic diarrhea** and a reason for failure to thrive in **infants in developing countries** (although rotavirus is more common), with some infections in bottle-fed infants in developed countries. EPEC has a **pathogenicity island (a somewhat foreign section in its DNA)** that codes for a type III secretion system; **intimin,** an **adhesin protein;** and several other *E. coli*–secreted proteins that are directly injected into the human cells. **EPEC attachment** (by **bundle-forming pili** and **intimin**) and subsequent **injection of proteins directly into epithelial cells cause actin rearrangement, which leads to small intestinal microvilli damage and surface effacement,** causing watery diarrhea.

c. **EHEC, of which the most notable strain is O157:H7,** may be found in food or water contaminated by **cattle feces** (mainly hamburger). Person-to-person transmission also occurs. **EHEC** (also called STEC [shiga toxin–producing *E. coli*] or VTEC [verotoxic *E. coli*]) causes attaching and effacing lesions, similar to EPEC. In the colon, however, EHEC **produces a hemorrhagic toxin called shiga toxin or verotoxin, which nicks the colon cells' 60S ribosomes, shutting down eukaryotic protein synthesis.** Clinically, this **noninvasive** infection is characterized by **frankly bloody diarrhea without prominent fever and usually without excess polymorphonuclear neutrophil leukocytes (PMNs). It may progress to hemolytic uremic syndrome (HUS) and acute renal failure. Antibiotics are currently contraindicated,** as most increase the production of the toxin and thus the risk of kidney damage.

d. **Enteroinvasive *E. coli* (EIEC) causes dysentery similar to that of shigellosis (fever, diarrhea, vomiting, abdominal cramping, and tenesmus).** Many patients have blood and pus in the stools, although some only have watery diarrhea. EIEC virulence is due to **invasion of intestinal epithelium.** Transmission may be associated with contaminated food.

4. **Neonatal septicemia and meningitis.** Strains of *E. coli* involved in meningitis are **K1-encapsulated strains** that are resistant to phagocytic removal from the bloodstream. (Note: Rates of neonatal meningitis with *Streptococcus agalactiae* or Group B Streptococcus, which was once more common than *E. coli,* has been declining due to better surveillance for Group A Streptococcus.)

III *Klebsiella*

A. KLEBSIELLAE are lactose-fermenting opportunists belonging to the Enterobacteriaceae. The most common species is *K. pneumoniae.*

B. *KLEBSIELLA* is an opportunist found in human **upper respiratory and gastrointestinal (GI) tracts.**

C. VIRULENCE. *Klebsiella's* large polysaccharide capsule is its major virulence factor. **Drug resistance is quite common.**

D. INFECTIONS

1. ***K. pneumoniae* occurs primarily in patients with underlying pulmonary disease or alcoholism.** (Note: *Streptococcus pneumoniae* is still the most common cause of pneumonia in alcoholics.) ***K. pneumoniae* tends to cause lobar pneumonia with difficult-to-treat abscesses and a dark-red bloody sputum** (currant jelly sputum). Sputum is usually not malodorous.
2. ***Klebsiella* urinary tract infections (UTIs)** occur most commonly **in patients with urinary catheters.**

IV *Shigella*

A. SHIGELLAE are nonmotile, non-lactose-fermenting Enterobacteriaceae that do not make hydrogen sulfide (H_2S). The most common species in the United States are *S. sonnei* and *S. flexneri.* The most severe is *S. dysenteriae* (rare in the United States).

B. VIRULENCE

1. All **shigellae invade through intestinal M cells, escape the phagosome, move laterally to adjoining cells by actin polymerization** (similar to *Listeria monocytogenes*), and induce apoptosis, thus creating the very characteristic shallow *Shigella* ulcers without vascular invasion (so bacteremia is rare.)
2. **Shiga toxin. Type 1 strains of *S. dysenteriae*** cause the **most severe disease,** because they also are the most potent **inducers of shiga toxin (an A-B component toxin), which cleaves the 60S ribosomal rRNA, inhibiting protein synthesis and damaging the intestinal epithelium and glomerular endothelium.**

C. RESERVOIR. The **only reservoir for shigellae is humans**—usually infants and children. Temporary fecal carriage after disease is common. The fecal-oral spread in day care is common due to the **extremely low infectious dose (10 to 100 shigellae).** Handwashing is critical to interrupt transmission.

D. INFECTIONS. Shigellosis symptoms vary widely, depending on the patient's nutritional status and age, the infective strain, and the dose. **Most cases are self-limiting.** Classically, this invasive disease manifests with **fever, tenesmus, frequent low-volume stools with both blood and pus, and abdominal cramping.**

E. LAB ID. Methylene blue stain of mucus from stools should show many PMNs typical of an invasive agent. Identification to genus and species is by **culture** on MacConkey and a specialized medium, such as salmonella/shigella agar, followed by chemical tests. If antibiotics are required, **susceptibility testing should be done,** due to the high level of multiple drug resistance.

V *Salmonella*

A. ***SALMONELLA*** is a highly motile, non-lactose-fermenting, encapsulated genus of Enterobacteriaceae. The more than 1,500 different serotypes are now classed as one species, *S. enterica,* with the serotype listed by antigen numbers. However, for communication ease, they are often still listed by the old serotype names. Of all the serotypes, *S. enterica* serotype Typhi is unique in having no nonhuman animal reservoirs.

B. ***SALMONELLA ENTERICA*** **SEROTYPE TYPHI** (a.k.a. *S. typhi* or Typhi)

1. **Virulence.** *S. typhi* attacks in the ileocecal region. In the submucosa, the polysaccharide capsule (Vi antigen) inhibits PMN uptake, but organisms still invade and kill intestinal M cells. Organisms are then picked up on the basolateral side by macrophages in the Peyer patches. The bacteria inhibit lysosome-phagosome fusion; they then replicate in the phagosome, kill the macrophages, and spread through the reticuloendothelial system, eventually reaching the bloodstream. Focal lesions occur in ~10% of cases. Eventually, *S. typhi* in the biliary tree is carried back into the intestinal tract.
2. **Reservoir.** ***Salmonella typhi's*** **only reservoir is humans. Transmission is fecal-oral.** Some people are **permanent carriers.** Infections are uncommon in developed countries (~500 cases per year in the United States). but worldwide estimates are of 17 million cases per year, with 600,000 deaths. Transmission is by the old "food, fingers, feces, and flies," as well as urine and raw shellfish from contaminated water.
3. **Infection. Typhoid symptoms** include **gradual onset of headache, loss of appetite, malaise, lethargy,** and **fever, with abdominal pain** and **constipation** early, **followed by a few short bouts of diarrhea after the second week.** If untreated, fever (up to 40°C [104°F]) and bacteremia may last several weeks. Hepatosplenomegaly and mesenteric lymphadenopathy are common. Rose spots appear on the abdomen in about 25% of cases, and focal lesions occur in the very young or old. Incubation period depends on dose. The disease varies from a prolonged, but mild, self-resolving fever to severe rupture of the spleen or intestines. Severity depends on previous exposure, inoculum, health status, age, and vaccination status. Vaccination is not routine in the United States.
4. **Lab ID. Early diagnosis is made by blood culture;** later diagnosis is made by blood, fecal, and urine cultures or a serum test (Widal) for patient antibodies to O and H antigens. Susceptibility testing is important, as drug resistance is becoming a problem.

C. ***SALMONELLA ENTERICA*** **SEROTYPE TYPHIMURIUM AND OTHER** ***SALMONELLA*** **SEROTYPES**

1. **Reservoirs.** There are **many animal reservoirs;** transmission is mainly from **raw poultry or eggs, reptilian pets,** or **raw milk.** Studies of milk have shown the infectious dose (30%, or ID_{30}) to be 10^5 organisms in people with normal stomach acid. Most likely to develop disease and to have serious disease are the very young or very old, oncology patients, immunocompromised patients, and people with reduced stomach acid. There is high potential for spread within families and institutionalized individuals.
2. **Infections**
   a. **Gastroenteritis.** *Salmonella* is invasive but usually self-limiting in the healthy, causing fever and producing an inflammatory but watery diarrhea with

abdominal pain, some nausea, and vomiting. **The most common agents—*S. enterica* serotypes Typhimurium and Enteritidis—usually do not invade vasculature to cause septicemias,** so focal lesions (seen at times with serotypes Dublin or Cholerasuis) are rare, except in sickle cell disease (SCD). Most salmonellae are resistant to serum killing.

b. **Osteomyelitis in SCD patients (not carriers).** Due to functional asplenism and defective opsonic activity and alternate complement pathways, **SCD patients have repeated infections** with **encapsulated organisms** and increased difficulty controlling Gram-negative bacteria, resulting in repeated septicemias and extremely high rates of osteomyelitis. **In SCD, *S. enterica* is the most common (>80%) causative agent of osteomyelitis.**

VI *Proteus*

A. *PROTEUS* spp. are opportunistic Enterobacteriaceae noted for **swarming motility** and **urease production.** *P. vulgaris* and *P. mirabilis* are common.

B. VIRULENCE FACTORS. *Proteus* grows well in urine and increases urine pH by urease production. This may result in a **strong ammonia smell in the urine** and may lead to the production of **renal calculi.** The swarming motility may aid entry to the bladder.

C. RESERVOIRS OF *PROTEUS* are water and human feces.

D. INFECTIONS. *P. mirabilis* causes UTIs, and *P. vulgaris* is an important nosocomial opportunist.

E. LAB ID. Swarming motility.

VII *Yersinia*

Yersinia, which belongs to the Enterobacteriaceae family, tends to be short and have bipolar staining when stained with Giemsa or Wayson stain. Pathogens include *Y. pestis, Y. enterocolitica,* and *Y. pseudotuberculosis.*

A. *YERSINIA PESTIS* CAUSES PLAGUE. *Y. pestis* has a **complex series of virulence factors**—some active in the flea vectors and others active at 37°C in the human or animal host.

1. **Epidemiology.** In the United States, the reservoir is **wild rodents in the Southwest desert. *Y. pestis* is transmitted to humans via the following:**
   a. **Flea bites**. When a flea takes a blood meal from a rodent that contains *Y. pestis,* the bacteria agglutinate, forming a blockage of the upper intestinal tract and creating an area for replication. However, because the flea's meal does not progress, the flea feeds again, regurgitating the yersiniae in the next mammal it feeds on.
   b. **Respiratory droplet from human cases with untreated pneumonic plague.** All patients should be considered highly infectious by respiratory route until after 72 hours of effective antibiotic therapy.
2. **Virulence factors.** In addition to chromosomal virulence factors and a chromosomal pathogenicity island, *Y. pestis* has three plasmids, two of which are not found in less virulent yersiniae. ***Y. pestis* has an affinity in the mammalian host for**

lymph nodes, but it has the **F1 capsular protein**, which prevents phagocytic uptake. There is a series of virulence factors that include ***Yersinia* outer membrane proteins (Yops) and a type III secretion system**, which allows the adhered organism to inject virulence factors directly into the human cells. This appears to be a defense system primarily aimed at killing phagocytic cells. *Y. pestis* produces a **plasminogen activator**, which appears to aid in the spread of the organism.

3. **Infection: Bubonic plague.** Symptoms include fever and painful, rapidly enlarging buboes (hemorrhagic suppurative lymphadenitis), fever, and conjunctivitis. Septicemia may lead to pulmonary emboli, pneumonic plague (highly contagious), and shock with disseminated intravascular coagulation (DIC). Fatality rate is high (50%) *if* untreated. **Pneumonic plague** presents with **fever, malaise, tightness in chest**, and a **pneumonia** that produces **first a mucoid and then bloody sputum.** It rapidly progresses to cyanosis and is fatal if treatment is delayed.
4. **Lab ID. Warn lab** of possibility! Stain bubo aspirates (Wayson and IF stain) by public health department; cultures are done, but treatment is not delayed for results.
5. **Prevention and treatment.** Isolate patient for 48 hours after antibiotics are started. Because ***Y. pestis* is facultative-intracellular**, antibiotics must be those that penetrate eukaryotic cells well and are effective against Gram-negative cells (usually gentamicin or streptomycin).

B. *YERSINIA ENTEROCOLITICA* is an enteropathogen that invades M cells over Peyer patches with variable symptomology. It causes **mesenteric adenitis** (mimics appendicitis), **enteric fever syndrome**, and a **reactive arthritis**, usually in adults.

1. **Characteristics.** Like *Listeria monocytogenes*, ***Y. enterocolitica* is a zoonotic, GI tract** organism that can **grow in cold.**
2. **Reservoirs and transmission.** *Y. enterocolitica* is transmitted by **direct animal contact** or in food products such as **raw milk.**
3. **Invasive enterocolitis.** Symptoms include **blood and pus in diarrhea, fever**, symptoms of appendicitis, or reactive polyarthritis. Because ***Y. enterocolitica* grows under refrigeration**, it may cause **blood transfusion–associated infections.**

VIII Other Enterobacteriaceae

Other genera of the Enterobacteriaceae family are seen as infective agents mainly in compromised or hospitalized patients. Those that cause septicemias include *Enterobacter, Citrobacter, Arizona, Providencia, Morganella*, and *Serratia* (noted for its salmon-red pigment).

Chapter 15

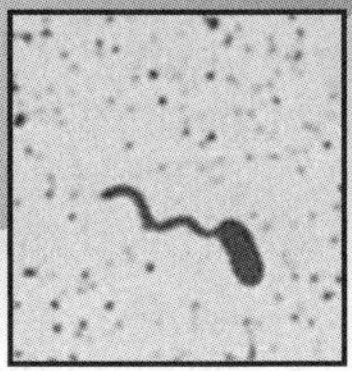
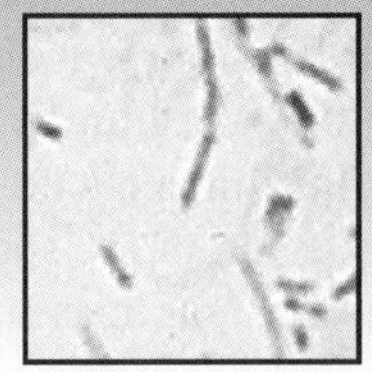
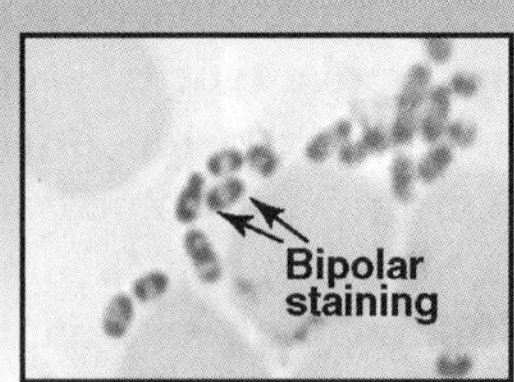

Gram-Negative Facultative Anaerobic Bacilli: Non-Enterobacteriaceae

Vibrio, Haemophilus, Pasteurella

I *Vibrio*

Vibrios are comma-shaped bacilli with one polar flagellum and are found in saltwater. Vibrios are facultative anaerobes, but, unlike the Enterobacteriaceae, they are oxidase-positive.

A. *VIBRIO CHOLERAE* is noninvasive and prefers an alkaline environment. O1 strains or O139 strains of the El Tor biotype of *V. cholerae* cause cholera.

1. **Epidemiology.** Human **fecal contamination of coastal waters** with *V. cholerae* leads to prolonged contamination and epidemics during plankton blooms. Water and shellfish may transmit. The infective dose is high if stomach acid is normal.
2. **Virulence.** *V. cholerae* regulates virulence factors, including toxin coregulated pili (TCP) and cholera toxin (CT) with upregulator (ToxR). **Cholera toxin binds to GM1-ganglioside receptors. The internalized A component ADP-ribosylates $G_{s\alpha}$, persistently stimulating cellular adenyl cyclase, which results in high cyclic adenosine monophosphate (cAMP) levels and massive efflux of ions and water into lumen,** primarily of the small intestine.
3. **Cholera.** *V. cholerae* strains cause a **profuse watery diarrhea with mucous flecks** ("rice-water stools"). Fluid loss is so great that **hypovolemic shock** will occur if electrolytes and fluids are not replaced. There may be vomiting, but there is no fever or blood in stools. There is a long-lasting **immunity to the serotype.**
4. **Lab ID.** Notify lab in nonendemic area, as a special **alkaline medium—TCBS** (thiosulfate citrate-bile salts-sucrose)—enhances growth. (Learn only TCBS/alkaline medium.)
5. **Treatment. Fluid and electrolyte replacement is critical.** Tetracycline reduces carriage and shortens the duration of illness.

B. *VIBRIO PARAHAEMOLYTICUS*

1. **Epidemiology.** ***V. parahaemolyticus*** is another coastal marine saltwater (halophilic) species that infects humans through **undercooked or raw sea food.**
2. **Disease.** *V. parahaemolyticus* **gastroenteritis** is characterized by a self-limiting, **explosive, watery diarrhea; vomiting; and fever. Diarrhea may be bloody.**

C. *VIBRIO VULNIFICUS*

1. **Epidemiology. *V. vulnificus*** is also halophilic, contaminating **oysters.** In the United States, some oyster beds along the **Gulf of Mexico coast** have been implicated.
2. **Diseases: Cellulitis, gastroenteritis, and septicemia. A serious necrotic cellulitis occurs in cuts (from shucking contaminated oysters). Ingestion of raw contaminated oysters causes a gastroenteritis; in patients with liver or iron-overload conditions, it may also cause septicemia and death.**

II *Haemophilus*

***Haemophilus* spp. are Gram-negative, pleomorphic, heme-loving rods.** These bacteria are fastidious, requiring growth factors from lysed blood. There are two important genera for humans: *H. influenzae* and *H. ducreyi.*

A. *HAEMOPHILUS INFLUENZAE*

1. **Epidemiology.** Nonencapsulated (a.k.a. **nontypeable**) **strains are part of our normal flora. Both nonencapsulated and encapsulated strains are transmitted by direct oral contact or respiratory droplets.**
2. **Virulence.** The **major virulence factor** for invasive disease is the **capsule.** In *unvaccinated* children <2 years old, ***H. influenzae* type b (Hib), which has a polyribitol phosphate capsule,** still commonly causes invasive disease.
3. **Diseases**
   a. **Purulent epidemic meningitis occurs in susceptible babies when Hib crosses the blood–brain barrier during septicemia.** Because of high vaccination rates in this country, *H. influenzae* is no longer the most common cause of meningitis in children <2 years of age. (*N. meningitidis* is more common.) **Monoarticular septic arthritis** is another possibility from septicemia with Hib.
   b. **Epiglottitis** is caused primarily by Hib. It occurs most often in 2- to 4-year-old boys. Symptoms include **fever, dysphagia, drooling, sore throat, inspiratory stridor,** and **cherry red epiglottis** protruding into the airway. Maintenance of airway and administration of antibiotics are important. There has been a dramatic decrease in cases due to vaccination.
   c. **Otitis media or sinusitis** is most commonly caused by **nontypeable strains of *H. influenzae.* Acute exacerbation of chronic bronchitis** (AECB) in smokers with chronic obstructive pulmonary disase (COPD) may involve a **new strain** of ***H. influenzae.***
4. **Lab ID. *H. influenzae* grows on chocolate but not blood agar.**
   a. **Chocolate agar** (a lysed blood agar) provides both the **X (protoporphyrin)** and the **V (NAD) factors required for growth of *H. influenzae.***
   b. **Satellite phenomenon.** If *Staphylococcus aureus* is also inoculated in the same area of a blood agar plate, **it produces NAD and lyses erythrocytes, releasing hemin. *H. influenzae* is able to grow as satellite colonies around *S. aureus.***
5. **Prevention.** Infant vaccine **(b capsular polysaccharide complexed to protein)** prevents Hib disease.

B. ***HAEMOPHILUS INFLUENZAE* VAR. *AEGYPTICUS*** **causes bacterial pink eye,** which has a purulent discharge.

C. ***HAEMOPHILUS DUCREYI*** is more common in the tropics and in the southern United States. It causes **chancroid,** a sexually transmitted disease (STD), characterized by **painful genital ulcers.** (Ⓜ You do cry with ducreyi.) **Chancroid lesions heal slowly and may increase the risk of HIV transmission.**

III *Pasteurella*

This genus consists of organisms that are **normal oral flora** in many animals. ***Pasteurella multocida*** is transmitted by dog, pig, and cat bites (domestic house cats or lions). **Cat bites** in particular need to be treated; **amoxicillin plus clavulanate** is the treatment of choice.

Chapter **16**

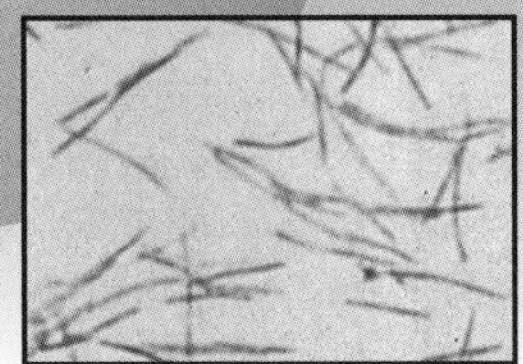

Gram-Negative Anaerobic Bacilli and Cocci

Bacteroides, Fusobacterium, Prevotella/Porphyromonas

I Gram-Negative Anaerobes

A. **RESERVOIRS.** Medically important anaerobic Gram-negative bacteria are opportunists found in the normal flora of the upper respiratory tract, mouth, colon, and/or female genital tract.

B. **DISEASES.** These bacteria cause **endogenous, mixed infections** (autoinfection), usually when they enter normally sterile, devitalized tissues or when they create anaerobic fluid pockets (empyema). They are major **causative agents of empyema, or abscesses** in respiratory, intestinal, or genital tracts or in bone or soft tissue. (*Staphylococcus aureus* may cause similar infections.) Anaerobes play a role in **chronic otitis media** and **sinusitis, aspiration pneumonia** (characterized by foul-smelling sputum), **decubitus ulcers,** and **soft-tissue infections in diabetic patients.**

C. **LAB ID.** Specimens must be **taken and transported anaerobically** (and quickly) to the lab for anaerobic culture. A syringe may be best.

II *Bacteroides*

Bacteroides is a genus of many species of **short, intestinal, Gram-negative bacilli** that are one of the most common intestinal bacteria and that are commonly involved in **intestinal abscesses (generally polymicrobic).** Most members of the *Bacteroides fragilis* group are aerotolerant (so not really fragile) and have an **antiphagocytic capsule,** which stimulates abscess formation, and a modified endotoxin with reduced toxicity. **Beta-lactamase production is common** in the *B. fragilis* group.

Prevotella and *Porphyromonas*

These are **normal oral and genital flora**, noted for melanin production, resulting in black colonies. (*Prevotella melaninogenica* was formerly known as *Bacteroides melaninogenicus*. The USMLE Step 1 usually gives you both names in recent name changes.) Infections with *Prevotella* or *Porphyromonas* include oral (and human bite), respiratory, and genitourinary tract infections. They are difficult to treat, requiring debridement as well as antibiotics.

Fusobacterium

Fusobacterium species are **Gram-negative anaerobes** of the **oral** and **gastrointestinal tracts.** Fusobacteria are **slender** cells with **tapered, pointed ends.** In addition to playing a role in intestinal abscesses, *Fusobacterium*, along with spirochetal oral flora, overgrows to cause Vincent angina, an ulcerative oropharyngeal disease.

Chapter **17**

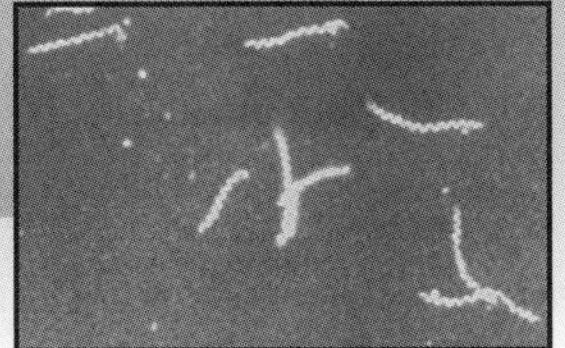

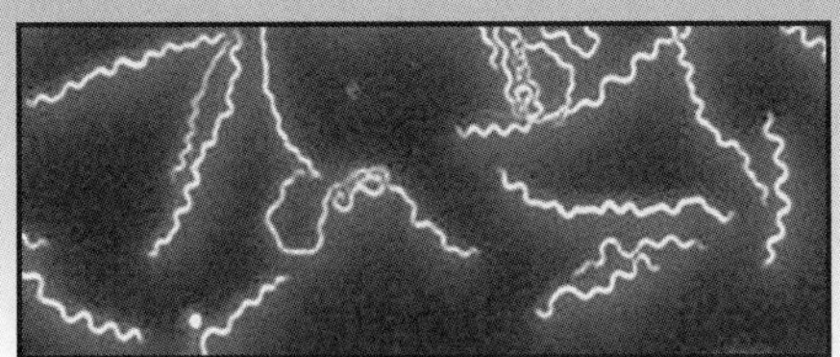

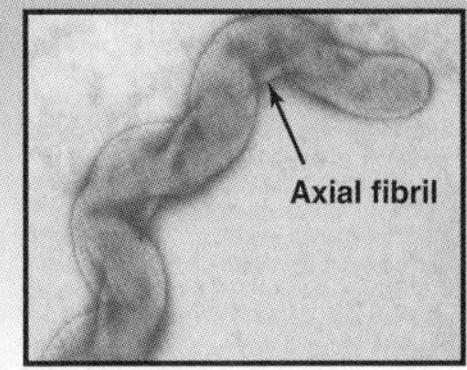

Spirochetes (Gram-Negative Envelope)

Treponema, Borrelia, Leptospira

I Spirochetes

Spirochetes are **Gram-negative, spiral-shaped bacteria** with **endoflagella** immediately under the outer membrane. These flagella (a.k.a. axial filaments) are wound around the body of the bacterium and are attached at each end, producing a **distinctive springing motility.** Spirochetes **do not show up well with Gram stain,** even though they have **Gram-negative peptidoglycan** and an outer membrane with endotoxin. They **are best visualized by dark field or fluorescent microscopy; *Borrelia* spp. show up with Giemsa.** Some are normal mucosal flora. Oral overgrowth of treponemes with fusobacteria in debilitated people causes trench mouth.

II *Treponema*

Treponema spp. **are thin, tightly coiled, microaerophilic, extracellular organisms.** Pathogenic strains, such as ***T. pallidum,* cannot reliably be cultured on inert media.** *Treponema* is easily visualized by **dark-field or immunofluorescent microscopy from the primary chancre.**

A. EPIDEMIOLOGY. Humans are the **only significant reservoir for *T. pallidum,*** which is transmitted **through sexual contact or across the placenta.**

B. INFECTION. Syphilis (untreated) is a multistage disease. Many features of syphilis are attributed to blood vessel involvement. The incubation period is variable—usually about 3 weeks. The following description is of *untreated* syphilis.

1. **Primary syphilis** consists of a **nontender, indurated ulcer (chancre)** with fairly smooth margins at the site of inoculation, with early bacteremia and spread to regional lymph nodes. **Chancres are highly infectious but heal spontaneously in 3 to 6 weeks.**
2. **Secondary syphilis** may appear 1 to 3 months later. *T. pallidum* spreads via the bloodstream, producing flulike disease with **hyperpigmented, but flat, cutaneous rash (infectious) and moist, infectious, papular perineal and genital lesions (*condylomata lata,* which are also infectious).** *T. pallidum* **may spread**

to any organ. Secondary syphilis may also "heal" spontaneously; however, about two thirds of patients do not clear the infection and eventually go on to latent syphilis.

3. **The latent stage** may be interrupted by relapses of the secondary syphilitic lesions and then become truly latent. Again, about two thirds of the untreated will progress to tertiary.
4. **Tertiary syphilis** may appear decades after the primary infection. **Aortitis is characteristic, along with variable central nervous system (CNS) involvement (neurosyphilis) or gummata in bones, skin, or viscera.**
5. **Congenital syphilis.** During the first 3 years of disease, ***T. pallidum* may cross the placenta.** Congenital syphilis symptoms range from **stillbirths to multiple fetal abnormalities; or babies may be asymptomatic until the age of 2 to 5 years.**

C. **LAB ID.** ***T. pallidum* is not cultured.** Diagnosis is based on clinical presentation and **microscopy and/or serology.**

1. **Microscopy.** *T. pallidum* is not seen by light microscopy, so **dark-field** (Figure 17–1) **or fluorescent microscopy** (with direct fluorescent antibody stain) are usually used in diagnosing primary syphilis. Negative microscopy does not rule out. Transudate from scrapings below the surface of primary lesions is used; avoid normal flora.
2. **Serological tests detect antibody. Two different antibodies are made in response to syphilis.** The earliest antibody is **antitreponemal antibody,** which is specific to treponemes. Approximately 1 week later, **nontreponemal antibody (a.k.a. reaginic ab or reagin)** appears. This nonspecific antibody appears to be

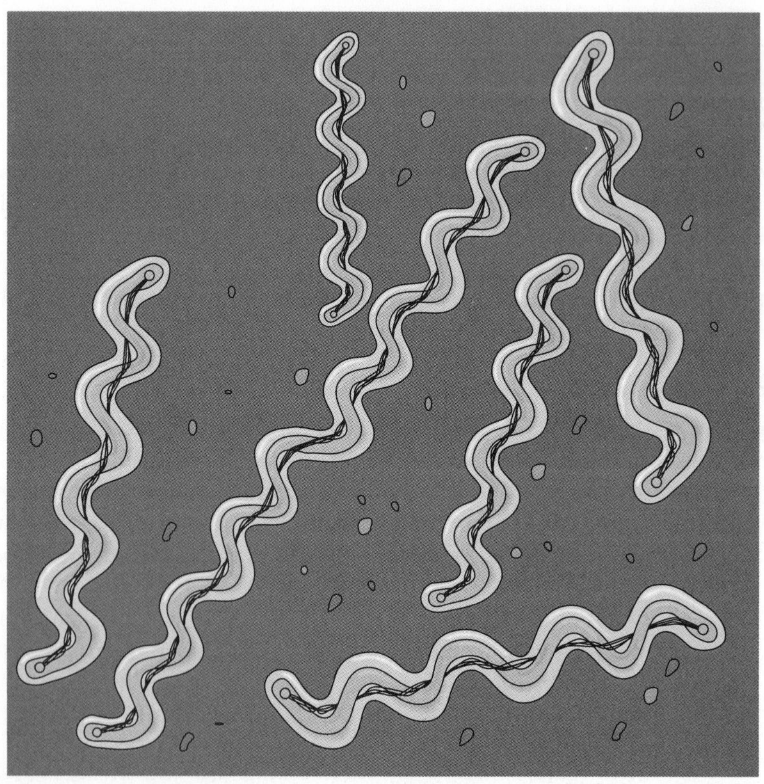

● **Figure 17-1** *Treponema pallidum* is seen in a "negative" pattern on dark-field microscopy.

TABLE 17-1 DIAGNOSTIC METHODS OF CHOICE FOR SYPHILIS

Diagnostic Method	Primary Syphilis	Secondary	Latent	Tertiary
Microscopy: Cannot rule out	Dark-field or DFA	Less common	–	–
Screening test (nontreponemal antibody): RPR, VDRL, ART*	± depending on timing; microscopy more reliable	+	+	±
Confirmatory test (treponemal antibody): FTA-ABS MHA-TP	± depending on timing	+	+	+

KEY: DFA, direct fluorescent antibody; RPR, rapid plasma reagin; VDRL, Venereal Disease Research Laboratory; ART, automated reagin test; FTA-ABS, fluorescent treponemal antibody absorption; MHA-TP, microhemagglutination–*Treponema pallidum.*

*Lupus and several other chronic diseases may give a positive VDRL, but genital lesions will not be present.

stimulated by cellular damage and may also be positive in autoimmune diseases, such as lupus or viral hepatitis. Nontreponemal antibody binds to mammalian cardiolipin (a very cheap antigen). These two antibodies are detected in the serologic diagnostic tests.

a. **Nontreponemal (reaginic) tests:**
   -are **inexpensive screening tests** (antigen = cow heart cardiolipin-lecithin)
   -detect reaginic antibody.
   -have high sensitivity (detect most cases) **but moderate specificity** (may be positive in other diseases too, but not other spirochete infections).
   -titers decrease with treatment or spontaneously in late latent phase.
   Nontreponemal tests include:
   i. **VDRL** (Venereal Disease Research Laboratory test)
   ii. **RPR** (rapid plasma reagin test)
   iii. **ART** (automated reagin test)
   If a nontreponemal test is positive, it must be confirmed with treponemal test.
b. **Treponemal antibody tests:**
   -use cultured treponemes or their antigens so **are expensive tests.**
   -detect specific treponemal antibody.
   -are used to confirm all positive nontreponemal screening tests.
   -have better specificity (except they may be positive in any other spirochete disease, such as Lyme disease).
   -antibody titer remains positive for many years, even with treatment.
   Common treponemal tests include:
   i. **FTAB-ABS** (fluorescent treponemal antibody absorption test)
   ii. **MHA-TP** (microhemagglutination–*Treponema pallidum* test)

See Table 17-1 below for which tests are used when.

D. TREATMENT. Treatment of primary or secondary syphilis infections is **benzathine penicillin.** A nontreponemal test will go down with treatment.

III *Borrelia*

Borrelia are **larger and more loosely coiled** than are treponemes. Known pathogenic species are arthropod-borne. **Antigenic variation is common.**

A. *BORRELIA BURGDORFERI*

1. **Virulence factors.** Because ***B. burgdorferi*** **may go intracellular,** polymerase chain reaction (PCR) from body fluids may not be positive. The organism **accesses immunologically privileged sites,** like CNS, tendons, vitreous humor, and synovia. **Antigenic variation occurs** (e.g., OspA, a major surface antigen of *B. burgdorferi* in ticks, decreases as the tick feeds on human blood, and OspC becomes the major surface antigen). **Endotoxinlike** compounds are found in the outer surface. **Peptidoglycan may play a role in causing joint inflammation. In chronic cases, there is some evidence of down-regulation of the host immune response.**
2. **Epidemiology. Reservoirs are white-footed mice and white-tailed deer.** Vectors are *Ixodes* ticks (a.k.a. black-legged or deer ticks). Nymph and adult ticks can carry the infection, but nymphs (0.5 mm) are the most common vector. ***Ixodes scapularis*** **is the vector in the northeast and north central United States.** ***Ixodes pacificus*** **is the vector in the western United States.**
3. **Lyme disease (LD).** Infection starts with an infected nymph or adult *Ixodes* tick bite. **Initial clinical symptoms may include erythema migrans (EM)—the spreading erythematous "target" lesion—as well as fever, chills, fatigue, headache, and muscle or joint pain. Dissemination (including to CNS) may occur as early as days to weeks later; multiple symptoms develop and may include multiple EM, Bell's palsy, fever, stiff neck, headache, limb numbness or pain, arrhythmias, persistent malaise, and fatigue. Late symptoms (polyarthritis, neurologic impairment, and fatigue) may develop months to years later, if the primary disease is not treated successfully.** Patients may have mixed infections with *Borrelia, Ehrlichia,* or *Babesia,* as *Ixodes* ticks may transmit all. Reinfections may occur. Congenital infections occur.
4. **Lab ID.** *Borrelia* may be seen in biopsy of advancing edge of lesion. Diagnosis is done by serology, PCR, microscopy (using acridine orange), or culture (modified Kelly medium). There are false negative serologies, possibly due to intracellular sequestering of organisms, heavy antigenic load (so no free antibody), antigenic variation of causative strains, and so on.
5. **Treatment** depends on the stage of the infection.

B. *BORRELIA RECURRENTIS* AND OTHER *BORRELIA* SPP.

1. **Epidemiology.** *B. recurrentis* is louse-borne. Other *Borrelia* infections are tick-borne.
2. **Relapsing fever** is characterized by an **abrupt onset of symptoms: shaking chills, fever, muscle aches, headache, delirium, cough, lethargy, splenomegaly, and hepatomegaly.** This is followed by **spontaneous resolution** of symptoms. **Symptoms recur and resolve again** about a week later, then recur less resolve one more time. **Antigenic phase variation** is responsible for recurring septicemias.
3. **Lab ID.** Diagnosis is by Giemsa stain. Serology does not work because of the antigenic variation.

 IV *Leptospira*

Leptospira is a genus of **fine spirochetes with hooked ends.** The most important species is ***Leptospira interrogans.***

A. EPIDEMIOLOGY.

Leptospires are found in a variety of **wild and domestic animals** and are transferred by **animal urine in water.** In the United States, the organism is usually transferred by contact with **dog, livestock, or rat urine (e.g., while working in sewers); by puddle stomping; or by recreating in contaminated water (e.g., jet skiing on rural lakes).**

B. **LEPTOSPIROSIS** may be subclinical but usually presents like influenza with or without abdominal pain, vomiting, and conjunctival suffusion. If untreated, it may progress to hepatitis, renal failure, and aseptic meningitis.

C. **LAB ID.** Leptospirosis is diagnosed serologically. Notify lab in potential cases.

Chapter 18

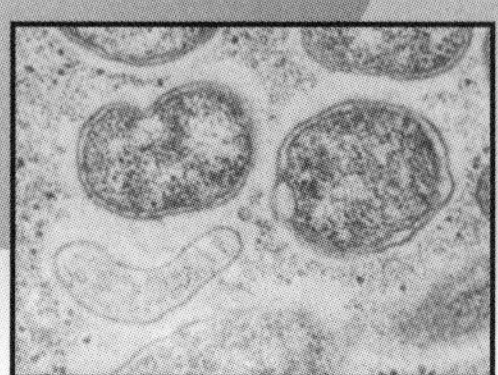

Rickettsiaceae and Relatives

Rickettsia, Ehrlichia, Bartonella, Coxiella

Despite recent taxonomic changes in the Rickettsiaceae family, which removed *Coxiella burnetii* and made some other name changes, the long-standing tradition of referring to the genera *Rickettsia, Ehrlichia, Bartonella,* and *Coxiella* as rickettsias continues. All but *Bartonella* are obligate intracellular parasites that make limited ATP. They have a Gram-negative cell envelope but are not seen well on Gram stain because of their small size. Known vectors and reservoir hosts for each rickettsia and disease are shown in Table 18-1. Where you need to know more than the vector and reservoir hosts, additional information is given in the text.

I *Rickettsia (Rickettsia rickettsii)*

A. ***RICKETTSIA RICKETTSII*** (Ⓜ the "double-named" rickettsia) causes the most important rickettsial disease in the United States—**Rocky Mountain spotted fever (RMSF). It is an obligate intracellular pathogen.**

B. **EPIDEMIOLOGY. Reservoir hosts are rodents, dogs, and Dermacentor ticks.** (The ticks are reservoirs because of **transovarian transmission.**) Vectors are *Dermacentor americanus* (wood), *Dermacentor variabilis* (dog), and *Amblyomma americanum* (Lone Star) ticks. Despite its name, RMSF occurs **most commonly in the eastern United States,** east of Oklahoma and Missouri (with North Carolina the highest). Tick and RMSF season is from April to September.

C. **TICK-INJECTED *R. RICKETTSII*** invade, replicate in, and kill endothelial cells of capillaries, causing vasculitis, which accounts for the disease's ease of spread and severity. Incubation is usually 2 to 6 days but can be up to 14 days. Common symptoms include **abrupt high fever, chills, headache, confusion, myalgia, nausea,** and **vomiting.** On the third to fifth day of symptoms, a macular **rash and swelling start on the ankles and wrists and spread to the trunk, palms, soles, and face.** The rash is present or detected in <60% of the cases. (Rash is more difficut to detect in dark-skinned individuals.) Mortality is up to 5% with treatment, 25% without.

TABLE 18-1 RICKETTSIAL DISEASES

Disease	Organism	Vector/Transmission	Reservoir
Rickettsial Diseases with Rash			
Rocky mountain spotted fever (rash starts distal, moves to trunk; + palms/soles)	***Rickettsia rickettsii***	**Dermacentor** or **Amblyoma ticks**	**Wild rodents, dogs; ticks also reservoir**
Rickettsialpox (lesions often mistaken for chicken pox)	***Rickettsia akari***	**Mite**	Rodents; usually in rodent-infested housing
Epidemic typhus (rash mainly on trunk)	*Rickettsia prowazekii*	Human body louse	Humans
Rickettsial Diseases Commonly with No Rash			
Ehrlichiosis (infection of monocytes or granulocytes)	*Ehrlichia chaffeensis* *Ehrlichia (Anaplasma) phagocytophilum*	Tick	Unknown
Cat scratch fever	*Bartonella henselae*	Scratch	Kittens
Bacillary angiomatosis, septicemia ± endocarditis	*Bartonella quintana* or *Bartonella henselae*	Human body louse in kitten scratch	Humans
Q fever*	*Coxiella burnetii*	Aerosols or dust inhalation; no vector†	Cattle, sheep, goats

*Rocky Mountain spotted fever and Q fever are currently the most important rickettsial diseases in the United States, but ehrlichiosis and bartonellosis are gaining importance.
†No significant arthropod vector in human disease.

D. **LAB ID.** Diagnosis of **RMSF** is by **immunofluorescent stain of skin biopsy, polymerase chain reaction (PCR) on blood**, or serology. (Note: Culturing *R. rickettsii* [done on tissue culture] is extremely hazardous, so is not routine.) The old Weil-Felix diagnostic test is based on the cross-reaction of *Rickettsia rickettsii* antibodies with OX antigens of *Proteus vulgaris.*

E. **TREATMENT.** Doxycycline (which penetrates human cells to kill the rickettsias) is started empircally any time RMSF is suspected.

II *Coxiella burnetii*

A. ***COXIELLA BURNETII*** **is a sturdy, Gram-negative, obligate intracellular pathogen (OIP).**

B. **VIRULENCE.** ***C. burnetii*** is **resistant to drying** and can survive in the environment. This bacterium grows in phagolysosomes, despite the presence of lysosomal enzymes, toxic radicals, and low pH.

C. **EPIDEMIOLOGY.** ***C. burnetii's reservoir is domestic livestock*** (particularly sheep), reaching **high titers in pregnant animals. Transmission is by inhalation of dust or aerosols of urine, feces, amniotic fluid, or placental tissue** or from unpasteurized milk or cheese from infected animals.

D. Q FEVER is an **atypical, community-acquired pneumonia**, presenting with **fever, headache, chills,** and a **mild dry hack.** Hepatitis often occurs. There is *no rash.*

E. LAB ID. Diagnosis is made by any of a number of serologic tests. Weil-Felix is negative. *Coxiella* is not seen well on Gram stain smear because it is very small. Specimens are hazardous; notify the lab.

III *Bartonella (Rochalimaea)*

A. BARTONELLA SPP. are pericellular (not intracellular) rickettsias. The two significant species are *B. quintana* and *B. henselae.*

B. INFECTIONS

1. **Septicemia with *B. henselae* or *B. quintana*** is seen in homeless, inner-city alcoholics. The vector is unknown. Diagnosis is by symptoms (**fever** with or without rash) and positive blood cultures. **Endocarditis** (culture negative by standard techniques) occurs in 20% of cases.
2. **In bacillary angiomatosis** (*B. henselae* or *B. quintana*), **vascular nodules** develop in HIV-positive individuals.
3. **Cat scratch fever** is caused by *B. henselae.* The organism is transmitted from an infected cat.

IV *Ehrlichia*

Ehrlichiae **infect human monocytes or granulocytes** and are seen as **mulberrylike clusters called *morulae.*** Ehrlichioses are **nonspecific febrile illnesses similar to RMSF, minus the rash** but with **leukopenia** and **thrombocytopenia.** Ehrlichiosis is **transmitted by ticks,** so coinfections with *B. burgdorferi* or *Babesia* may occur.

Chapter **19**

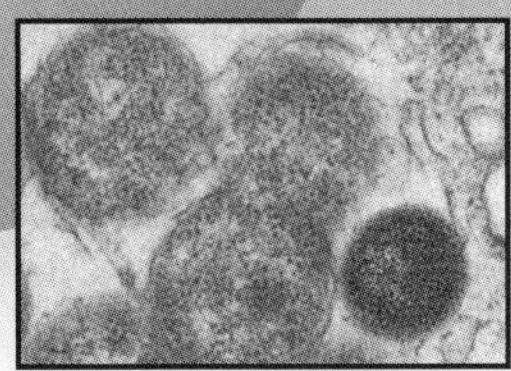

Chlamydiae

Chlamydia

I General Characteristics

A. CHLAMYDIAE are **obligate intracellular bacteria** that cannot make ATP or many amino acids. Chlamydiae do not have peptidoglycan; they do have an outer and inner membrane. They are not seen well on Gram stain.

B. THE LIFE CYCLE of chlamydiae is complex:

1. The **infectious, extracellular** (Ⓜ out in the "elements") form, called the ***elementary body*** (Figure 19-1), is **resistant to drying,** is **metabolically inactive,** and has ligands that can bind to epithelial cell receptors, stimulating endocytic uptake.
2. The **intracellular** form is known as the ***reticulate body.*** Endocytosed elementary bodies develop into **metabolically active, dividing reticulate bodies** that produce new elementary bodies. Aggregates of reticulate bodies in the endosome (inclusion bodies) can be visualized with fluorescent antibody stains.

II Virulence

All chlamydiae invade epithelium. Damage results from **granuloma formation,** which may lead to serotype-specific consequences, such as **fallopian tube blockage (serotypes B, Ba, D-K), corneal scarring (A, B, Ba, C),** and **lymphatic blockage (L1, L2, L3).**

III *Chlamydia trachomatis*

Chlamydia trachomatis has numerous serovars (strains) that cause three major diseases. Each group of serovars has a distinct epidemiology and disease pattern.

A. SEROTYPES D-K, B, BA are common in the United States and cause **reproductive tract infections, neonatal pneumonia,** and **inclusion conjunctivitis** in all economic classes.

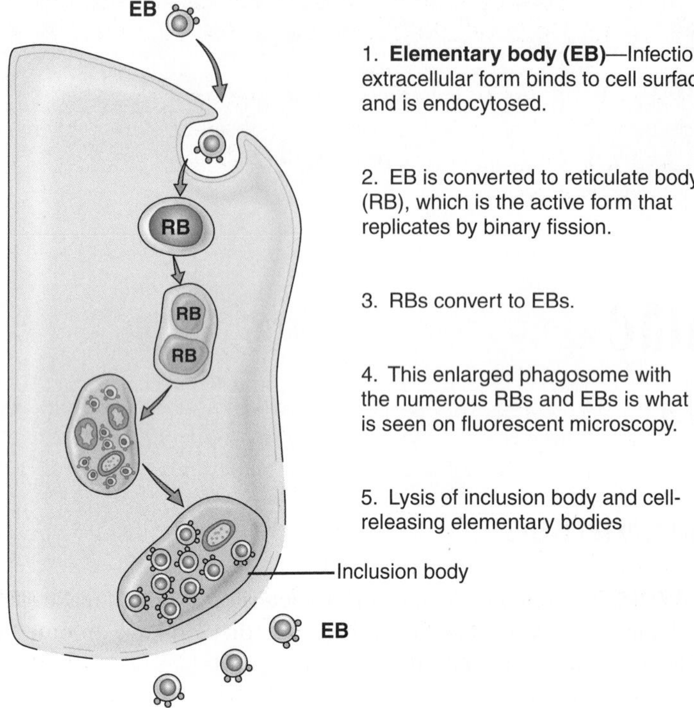

EB
RB
RB
RB
1. Elementary body (EB)—Infectious extracellular form binds to cell surface and is endocytosed.
2. EB is converted to reticulate body (RB), which is the active form that replicates by binary fission.
3. RBs convert to EBs.
4. This enlarged phagosome with the numerous RBs and EBs is what is seen on fluorescent microscopy.
5. Lysis of inclusion body and cell-releasing elementary bodies
Inclusion body
EB

● **Figure 19-1** Life cycle of chlamydiae.

1. **Infection in adults, especially young sexually active adults.** *C. trachomatis* serotypes D-K are transmitted sexually or by other direct contact. Infections are often subclinical but may still cause significant fallopian tube damage. Repeated infections may occur; infertility increases to 50% with three or more infections. Chlamydial genital infections (about 5×10^5 per year) are more common than gonorrhea (about 3×10^5 per year). Chlamydial infection increases risk of acquiring HIV.
2. **Infections in neonates** (acquired during delivery)
   a. **Chlamydial inclusion conjunctivitis.** Infection is micropurulent, with onset of symptoms in the neonate between days 8 and 10. Erythromycin eye ointment (used at birth to prevent *N. gonorrhoeae* infection) does not prevent chlamydial conjunctivitis. Systemic erythromycin must be used to treat the eye infection and prevent chlamydial pneumonia.
   b. **Chlamydial pneumonia (interstitial).** Onset usually occurs from 2 or 3 weeks after birth to 6 months of age. Symptoms start with **rhinitis** and develop into a **distinctive staccato cough.** Babies are often afebrile but have difficulty feeding. Infections may be associated with decreased pulmonary function later in life.
3. **Lab ID.** Nucleic acid amplification or **probe tests** are used for genital specimens. **Inclusion bodies may** be identified from direct specimens or tissue cultures with fluorescent antibody (or iodine) staining. Cultures, when done, are grown in McCoy cells.

B. SEROTYPES A, B, BA, AND C cause trachoma, a major **cause of blindness** in Asia and Africa (10 to 20 million people). Trachoma is also endemic among Native Americans in the southwestern United States. In trachoma, **chlamydiae invade the epithelium of**

the conjunctiva, causing chronic follicular keratoconjunctivitis. Follicular scarring leads to inturned eyelashes, corneal scarring, and, ultimately, blindness.

C. **SEROTYPES L1, L2, AND L3** cause **lymphogranuloma venereum.** This sexually transmitted disease, prevalent in Africa, Asia, and South America, presents with a **painless primary lesion, fever, headache, and myalgia.** Secondary symptoms include **inflammation and swelling of lymph nodes,** with systemic spread. **Tertiary symptoms are ulcers, fistulas,** and **genital elephantiasis.**

IV *Chlamydia psittaci*

A. **RESERVOIR AND TRANSMISSION.** This **zoonotic species of *Chlamydia*** consists of obligate intracellular parasites that invade respiratory epithelium to cause **pneumonia in birds (pet birds, zoo birds, turkeys) and humans.** Transmission to humans is by exposure to an infected bird or to dried bird excrement.

B. **PSITTACOSIS** is an **atypical pneumonia.** Early symptoms include **headache, high fever and chills, malaise, anorexia, myalgia, arthralgia, and pale macular rash.** (Remember, these are Gram-negative bacteria.) Pulmonary symptoms are a nonproductive cough, rales, and consolidation. Central nervous system involvement is common, usually manifesting as headache, but in severe cases, as encephalitis.

C. **LAB ID.** *C. psittaci* is usually diagnosed by serology; it is not seen with iodine staining.

V *Chlamydia pneumoniae*

Chlamydia pneumoniae is a **human pathogen** spread by **respiratory droplets.** Infection is **common but usually not severe.** Symptoms include **bronchitis, pneumonia,** or **sinusitis.** There are an estimated 200,000 to 300,000 cases per year, mainly in adults 18 to 45 years of age. *C. pneumoniae* may be associated with atherosclerosis. Laboratory identification, when needed, is by serology. Treatment, when needed, is usually with azithromycin.

Part III

Viruses

Chapter 20

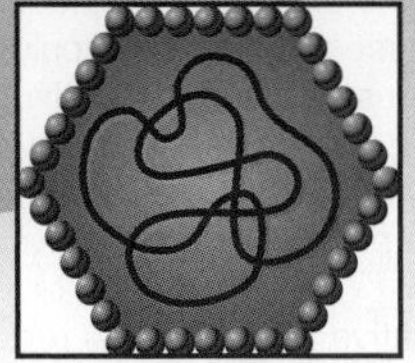
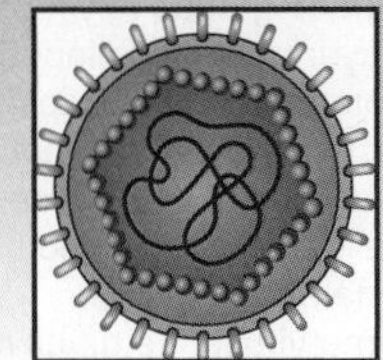
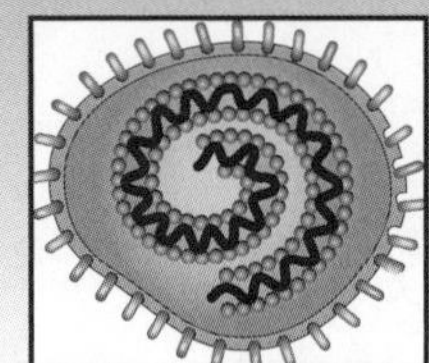

Viral Basics

I Exam Strategies

Viruses are important on the USMLE Step 1 exam. First, understand the big picture of how viruses infect a cell (virus binding through viral ligands to host cell receptors; this is responsible for both host and tissue specificity). Viruses, as obligate intracellular parasites, must replicate in the host cell, and because of their small size, they use many host cell components, including enzymes, ribosomes, and ATP. You then need to understand the big picture of how a DNA virus, a positive (+) strand RNA virus, a retrovirus, and a negative (–) strand RNA virus each replicate. You should also learn to recognize the basic presenting symptoms of the major viral diseases and know the causative agent, and then learn the family of each virus and the basic characteristics of each family.

Many Step 1 questions start with a clinical scenario requiring you to recognize the disease and know the causative agent. Those questions then ask major basic science questions (e.g., about replication intermediate or alcohol susceptibility) or they list the causative agents by description (rather than by name), such as "a ds DNA virus that makes its own lipid coating and replicates entirely in the cytoplasm," which you should know is a poxvirus if you use the above strategy.*

II Viral Structure

Viruses are **acellular, obligate intracellular organisms.**

A. VIRIONS are the **mature, released viruses,** which can infect and take over the machinery of a host cell to make more virus.

B. VIRAL GENOMES are either DNA or RNA. This **nucleic acid (NA) codes for all viral enzymes and structural components, except for the host-derived membrane component of enveloped viruses. Some RNA viruses have more than one viral chromosome:** HIV is diploid (two identical chromosomes); others, like influenza, are segmented, with each of the eight pieces of RNA carrying the genetic material for a gene.

*The following abbreviations will be used for single stranded: ss, (–) strand, or (+) strand; for double stranded, ds will be used.

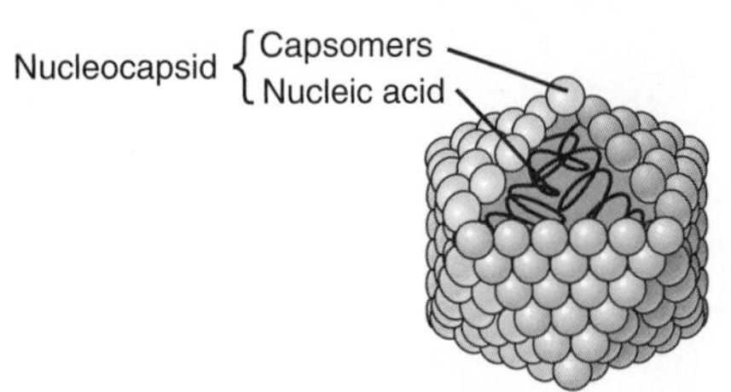

High-Yield Concepts: Virions

1. Surface groups (glycoproteins embedded in the lipid bilayer of the enveloped viruses or surface capsid proteins of naked capsid viruses) bind to specific host cell receptors, inducing uptake and thus determining host specificity and tissue tropism.
2. The nucleic acid directs the production of the next generation of virus.
3. Some viruses (e.g., all negative RNA viruses) require a virion-associated polymerase (the actual protein in the nucleocapsid) to be infectious.

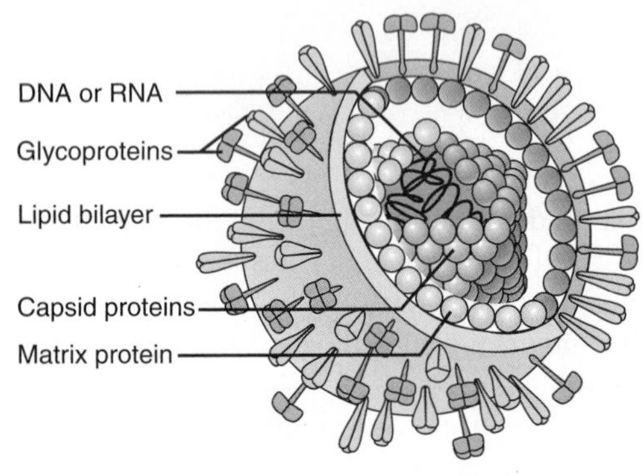

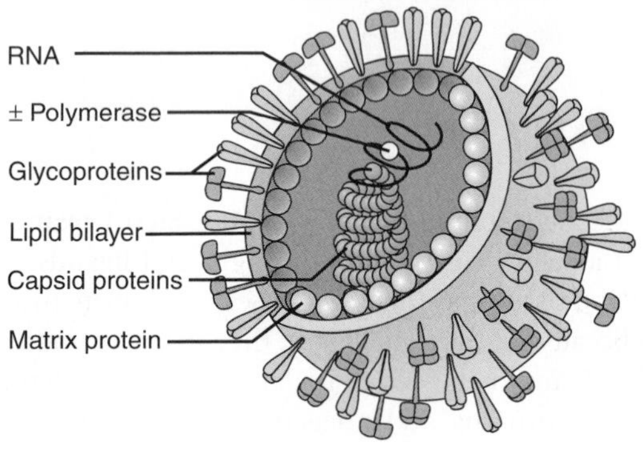

● **Figure 20-1** Virion (the mature virus) structure.

C. **VIRION PROTEINS include the following:**
   1. **Structural proteins,** such as **capsomers, which make up the viral capsid** or "head," and **glycoproteins (e.g., hemagglutinins),** which are incorporated into the envelope of enveloped viruses and are **used in binding to human cells.**
   2. **Enzymes. All (–) RNA viruses, plus the retroviruses, hepatitis B virus, and Poxviridae, have some type of polymerase protein in the virion,** without which they would not be able to replicate.

D. **GENOMIC NUCLEIC ACID, with proteins such as polymerases and capsid structural proteins, make up the viral nucleocapsid.** Nucleocapsids may be icosahedral or helical.

E. **VIRUSES** may be naked or enveloped (Figure 20-1).
   1. **Naked viruses** have **no lipid envelope;** instead, the outer component is the tight **icosahedral** nucleocapsid. Naked viruses are **not easily damaged by organic solvents** or detergents (bile salts) and are **more resistant to killing by chlorination** than are enveloped viruses. Naked viruses assemble large numbers of progeny inside the cell, which are all **released at one time by cell lysis.** This release immediately kills the host cell. Thus, naked viruses do not release small quantities of virus over a long period of time, as occurs with some enveloped viruses. Ⓜ Viral families that are **naked viruses are "The naked PAPs PiC Reo": Parvoviridae, Adenoviridae, and Papovaviridae (all DNA); Picornaviridae and Caliciviridae (both (+) RNA); and the ds RNA Reoviridae.** (Note: None of the (–) RNA viruses is naked.)

2. **Enveloped viruses** direct synthesis of specific viral glycoproteins that are inserted into their host's membrane. **The virus buds out of the cell, taking on the modified host's membrane as an envelope.** This "budding" type of viral release **does not kill the cells outright.** Thus, some enveloped viruses can set up a **chronic productive infection in which virus is produced over time from the same cells.** Ultimately, the cell will die due to the "viral takeover" or the triggering of apoptosis. Some enveloped viruses, such as HIV, are actually cytotoxic. In addition, viral glycoproteins inserted into a human cytoplasmic membrane **trigger a cellular mediated immune response, attacking the infected human cells,** which is often the cause of cell death and limits infection.

F. **HOST CELL RANGE** (e.g., what type of animal is infected) and **tissue tropism** (i.e., what tissues are infected within that host) **is determined by specific surface proteins or glycoproteins on each virus.** These proteins must bind to specific cellular receptors.
1. **Naked icosahedral viruses bind to cell receptors through specific surface proteins of the capsid.**
2. **Enveloped viruses bind to specific host cell receptors through viral glycoproteins embedded in the envelope.** Organic solvents inactivate enveloped viruses by damaging the lipid envelope, causing the critical cell binding surface glycoproteins to "fall off." (There are no backup cell-binding ligands on the nucleocapsids underneath the envelope.)

Viral Replication

Viruses take over their host cells and then use both host- and virus-specified components to make new virions.

A. **INFECTION STARTS** with viral **attachment** through viral surface ligands to specific receptors on the human cell membrane; **this stimulates viral penetration through either of the following:**
1. **Fusion of the membrane with the viral envelope** (as in HIV, Poxviridae, or Herpesviridae)
2. Stimulation of **pinocytosis** (all naked viruses)

B. **EARLY MACROMOLECULAR SYNTHESIS.** Once inside the cell, the viral nucleic acid is released and transported to where it will ultimately be duplicated. The first critical function is to produce **early mRNA and the proteins required to convert the cell into a viral factory.** Figure 20-2 shows how each major type of virus produces early mRNA. The viral proteins produced from these early mRNAs carry out a variety of mechanisms to take over the host cell, often causing cytopathology and, eventually, killing most cells except those latently infected or transformed.

C. **REPLICATION OF GENOME.** After early proteins are made, the virus can take over and replicate its own genome. Although the nucleic acid of the progeny virus should be identical to that of the parent, there is a high mutation rate because viral polymerase's editing of mistakes range from poor (e.g., DNA viruses with their own polymerases, as in herpes simplex virus [HSV]) to none (RNA viruses). The pattern of replication for major viruses is shown in Table 20-1. (Note: For all ss RNA viruses [except retroviruses], a homologous replicative RNA intermediate is required.)

Production of Viral mRNA

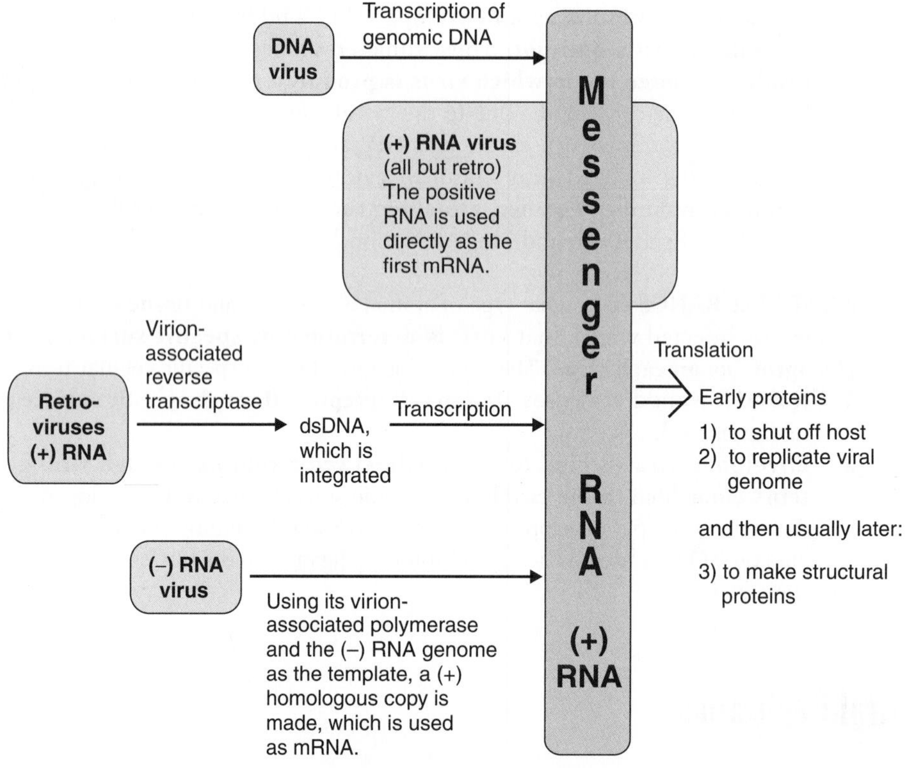

● **Figure 20-2** Scheme by which various viral groups make their early viral mRNA.

TABLE 20-1 SYNTHESIS OF NEW VIRAL GENOMIC NUCLEIC ACID

Genome	Replication	New Genomes (same as original)
ds DNA	DNA separates, with each half serving as the template to produce new double strands.	ds DNA
(+) RNA (except retroviruses)	Makes a (–) homologous strand (using its RNA-dependent RNA polymerase, which it has made as one of its early proteins). The (–) homolog then serves as the template to make more (+) RNA.	(+) RNA
Retroviruses (+) RNA	Using virion-associated reverse transcriptase, a ds DNA is made and integrated into the host cell's chromosome. Intact RNA transcripts serve as new genomes.	(+) RNA (retroviral)
(–) RNA	Using the virion-associated, RNA-dependent RNA polymerase, a (+) replicative intermediate is made that is used to make more (–) RNA.	(–) RNA

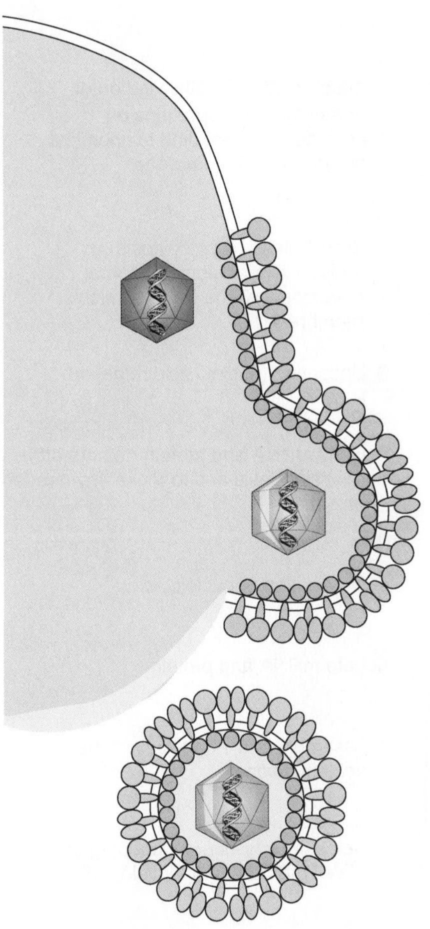

1. Enveloped viruses synthesize proteins, which are then glycosylated and inserted either in internal membranes (e.g., nuclear) or in the plasma membrane, depending on the specific virus.
2. The viral capsids accumulate by the membranes containing these glycosylated proteins and matrix proteins. They then bud through these membranes, wrapping themselves in the lipoprotein bilayer modified with these glycoproteins, which are essential for binding to the host receptor to start the next round of infection. (If the virus is hit with an organic solvent, the membrane is damaged and the glycosylated proteins drop out. The virus can no longer bind to trigger uptake.)
3. Mature virion is released without lysing the cell. This permits a cell to produce virus over a longer time.

● **Figure 20-3** Viral encapsulation. All surface glycoproteins added to the membrane (shaded in the figure) are viral coded.

D. VIRAL ASSEMBLY. Ultimately, all viral proteins are made, and the virus is assembled.

E. VIRAL RELEASE. Cytopathic viruses kill the host cell; most release virus by lysis. (Most naked viruses lyse cells, as shown in Figure 20-3.) Noncytopathic viruses (enveloped viruses) may be produced over a fairly long time and released by budding through the host membrane. Those that bud through other membranes, similar to Golgi, may exit by a reverse pinocytosis. Some enveloped **viruses with fusion proteins may enter an adjoining cell without re-entering the extracellular spaces** by fusing the two cell membranes, thus creating **giant cell syncytia** and avoiding the extracellular environment. However, because these viruses alter host cells, taking over the normal essential cellular processes to make new virus, most cells that produce virus will eventually die.

F. FIGURES USED IN TESTING. Steps in a **generalized infection**, including the **eclipse** and the **latent periods**, are shown in **Figure 20-4.** The **one-step growth curve** shown in **Figure 20-5** has been critical to our basic understanding of viral replication and is used to test your knowledge of replication. When this test was first run, only intact viruses could be detected. The diagram is highly generalized. You must be able to identify the eclipse, latent, and intracellular accumulation periods, which are labeled in the legend.

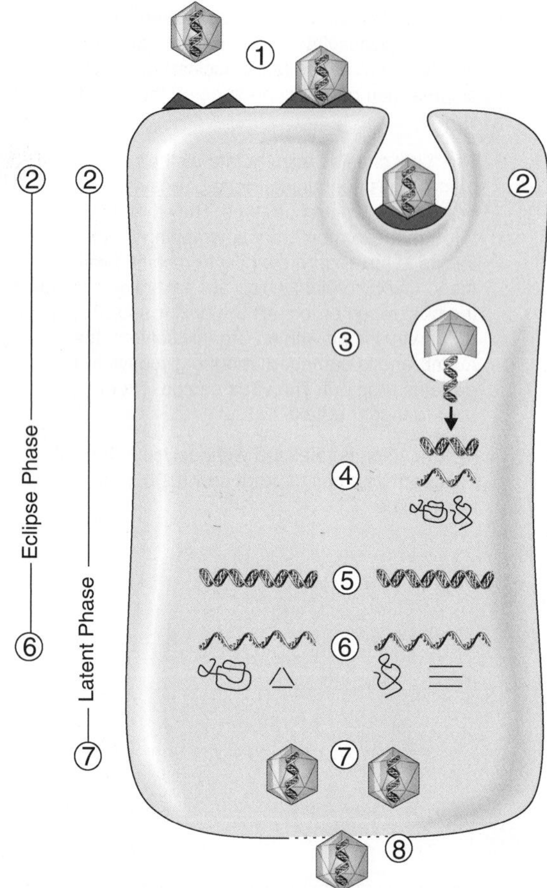
Steps in Generalized Viral Infection
1. Attachment—Specific viral outer proteins (or glycoproteins on envelope viruses) bind to chemical groups on cell membrane.
2. Virus uptake by pinocytosis (as shown) or by fusion of the viral envelope with the cytoplasmic membrane
3. Uncoating (nucleic acid released)
4. Early mRNA and protein (to shut off host synthesis and to make any needed enzymes)
5. Duplication of nucleic acid
6. Late mRNA and protein
7. Assembly and intracellular virus accumulation
8. Release by lysis or by budding out of cell membrane (if enveloped)
Eclipse Phase
Latent Phase
Stages 2–6 Eclipse Phase—no internal or external virus
Stages 2–7 LatentPhase—no external virus

● **Figure 20-4** Viral replication overview of a generalized infection. The eclipse phase is the time from uptake of the virus to just before the assembly of the first intracellular virus (stages 2–6). The latent phase is the time from the initial infection to just prior to the first release of the extracellular virus (stages 2–7).

IV Host Defenses and Viruses

Each virus has a complex relationship with the human host. Each portal of entry to the body has different extracellular physical and chemical barriers, and each virus initially reacts according to its surface components. For example, naked viruses, such as enteroviruses, are much sturdier in terms of surviving in the bile salts of the gastrointestinal tract. Viruses must be able to evade (or actually moderate) the host defenses. If the virus enters an already immune host with high levels of neutralization antibody, the infection will be aborted.

A. INNATE IMMUNITY. In an immunologically naive host, viruses attach to the host cells and infect. However, **natural killer (NK) cells** identify viral-infected human cells, triggering apoptosis in an attempt to kill the cell without producing virus.Those viruses

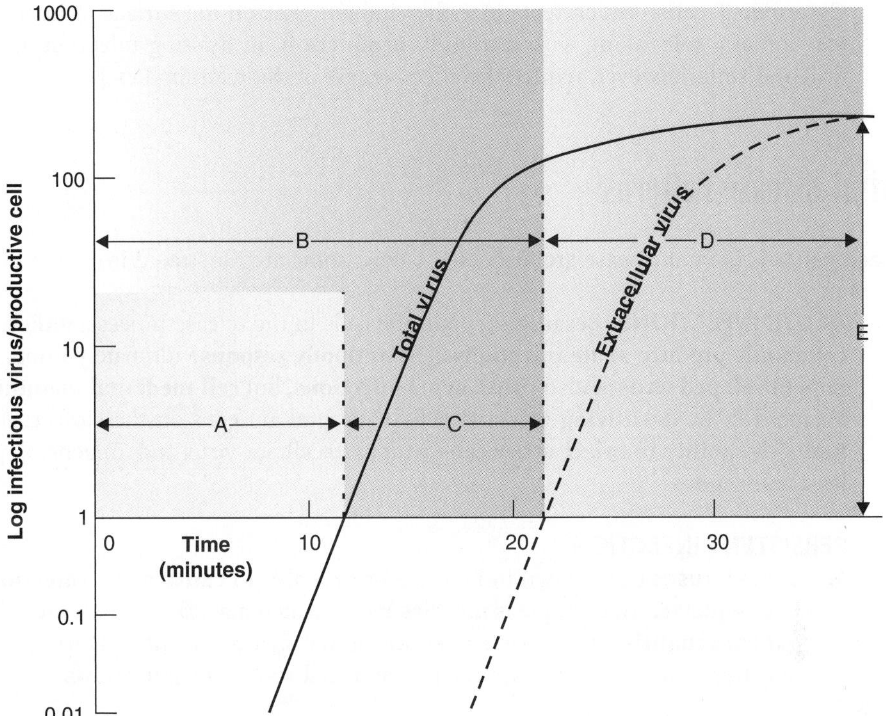

● **Figure 20-5** The one-step viral growth curve. Cells were infected; the remaining free virus was removed. Only complete, assembled virus was detected. At each time point, two samples were taken: (1) Supernatant (no cells) was assayed for free extracellular virus. (2) The entire sample (media and cells) was lysed and assayed for total virus, which included intra- and extracellular virus. **(A)** Eclipse phase. **(B)** Latent phase. **(C)** Intracellular accumulation. **(D)** Rise phase (time it takes to release all the virus). **(E)** Viral yield (amount of virus per cell).

are able to slow apoptosis and turn on the production of needed host enzymes or produce their own. They can, thus, successfully replicate before apoptosis kills the cells. Another part of the innate immune system is the production of **interferons** in viral-infected cells. **Interferons** are **most strongly triggered by ds RNA** (replication intermediates) and have the ability to **shut off virus *production*** in virus-infected cells. (Interferons are discussed, along with more details on the immune response to viral infection, in Johnson AG. *High-Yield Immunology*. Philadelphia, PA: Lippincott Williams and Wilkins; 2002.)

B. ADAPTIVE IMMUNITY. The adaptive immune response plays a major role in resolving infections and preventing reinfection. These interactions are complex, and the outcome depends on the host immune status, the virus, the route and dose of the virus, and the virus's ability to modify the immune system. Some viruses directly infect cells of the immune system. Many viral gene products will try to modulate the immune system so it is less effective in allowing virus replication. Measles virus infects respiratory mucosa and reticuloendothelial cells causing mucosal damage and a 1–2-week immunosuppression, which leads to a high death rate from additional infections in malnourished kids.There are also viruses that interfere with antigen presentation. Many viral rashes are a result of a delayed hypersensitivity to virus in the skin following a secondary viremia.

Cytotoxic T cells (which recognize the viral antigens on the surface of infected cells) play a major role, along with **antibody production,** in limiting infection. (Antigenic drift and shift, however, reduce the effectiveness of these antibodies.)

V Viral Disease Patterns

Generalized patterns of viral disease are discussed below; some are illustrated in Figure 20-6.

A. ACUTE INFECTIONS. Because they kill the cells in the release process, **naked viruses commonly produce acute infections.** The **antibody response** ultimately **limits reinfection. Enveloped viruses also cause acute infections,** but **cell-mediated immunity** plays a larger role by **destroying infected cells** with viral antigens on the surface. Antibody **limits the ability to infect other cells** with extracellular virus and, in general, prevents later reinfections.

B. PERSISTENT INFECTION

1. **Some viruses** (not shown in Figure 20-6) are able to **cause acute infections with late sequelae.** An example is **measles virus,** which may cause **subacute sclerosing panencephalitis** many years after the initial acute infection. During the acute infection, viral immunosuppression and cell-to-cell fusion allows the virus to

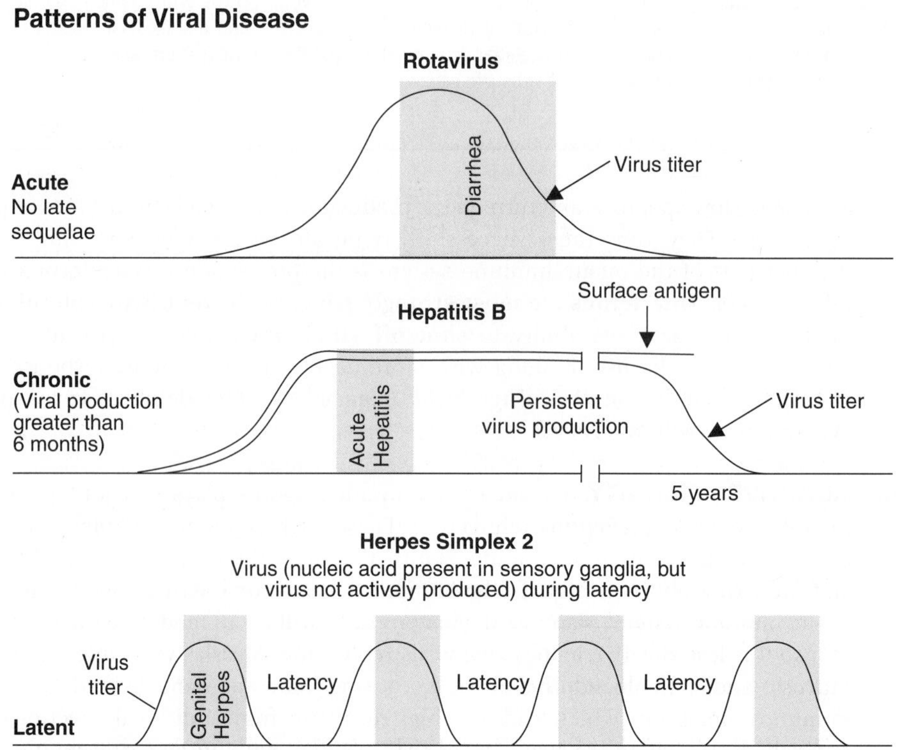

● **Figure 20-6** Patterns of viral disease.

spread to the central nervous system. Although nucleoproteins may be seen in the electron microscope, infectious virus cannot be cultured, and envelope proteins are not made. The virus does not bud out, but it is still able to move to new cells, probably by cell-to-cell fusion.

2. **Chronic infections.** These productive infections (with long-lasting production of virus) are possible with enveloped viruses, such as hepatitis. *Without* antiretroviral drugs, **HIV production continues from a single cell at a low level for years.** Therefore, HIV is not considered latent.
3. **Latent infections.** Latent infections (with **viral nucleic acid present** in a cell but **not actively producing virus,** except under certain conditions) occur with some viruses, such as herpes simplex.

VI Characteristics of Hepatitis Virus and Disease

The following chapters will discuss DNA viruses, (+) RNA viruses and retroviruses, (–) RNA viruses, and ds RNA viruses. Before moving on to that discussion, however, a comparison of the diverse viruses that cause hepatitis is useful in preparation for Step 1.

A. **ACUTE SYMPTOMS OF HEPATITIS** usually include **fever, malaise, headache, dark urine, vomiting, and jaundice.**

B. **HEPATITIS VIRUS A, C, D, and E are RNA viruses. Hepatitis B, however, is a DNA virus that replicates through an RNA intermediate.** (Ⓜ Why would a self-respecting DNA virus replicate through an RNA intermediate? Peer pressure to replicate through an RNA intermediate, like all of the RNA hepatitis viruses do!?)

C. **FECAL-ORAL TRANSMISSION occurs only with the naked hepatitis viruses** (A and E), which are not damaged by enteric detergents (bile salts). (Note: Fecal virus, which is present for about 6 weeks in hepatitis A, can be transmitted by oral-anal sex during that period. Also, because there is a short viremia—about 2 weeks with hepatitis A—both A and E may be sexually transmitted. Neither hepatitis A nor hepatitis E sets up chronic infections, because infected cells make a lot of virus and kill the cell to release that virus. Eventually, antibody plays a major role at preventing new cells from being infected.

D. **THE ENVELOPED HEPATITIS VIRUSES** are all **blood-borne,** may be transmitted **sexually,** and have the potential to become **chronic. Chronic hepatitis increases the risk of primary hepatocellular carcinoma.** Infants acquiring hepatitis B perinatally, if untreated, are most likely to develop chronic hepatitis B infection. Although acute hepatitis C is often asymptomatic, it has a very high rate of chronicity.

E. **HEPATITIS D** is a **defective RNA virus** that **requires hepatitis B surface antigen** in order to replicate. **Coinfections (B and D acquired together) are serious. Superinfections (i.e., when a person already has hepatitis B and then acquires hepatitis B again with D) are even more likely to be fatal.**

F. **THE FIVE MOST COMMON HEPATITIS VIRUSES** and their diseases are shown in Table 20-2. Each virus will be discussed later. Hepatitis G, not shown, is similar to C.

TABLE 20-2 COMPARISON OF HEPATITIS VIRUSES

Name	Viral Characteristics	Distinguishing Features
Hepatitis A "Infectious" hepatitis	**Naked** icosahedral (+) ss RNA: Picornavirus	Fecal-oral, often food-borne; Onset usually abrupt, disease mild; 0.5% mortality No chronic infections; no association with cancer
Hepatitis E "Enteric" hepatitis	**Naked** icosahedral (+) ss RNA: Calicivirus	Fecal-oral, often food-borne Normal people: onset abrupt, disease mild; 1%–2% mortality Pregnant patient: severe; 20% mortality No chronic infections; no association with cancer
Hepatitis B "Serum" hepatitis	**Enveloped,** partially ds DNA Replicates through an RNA intermediate: Hepadnavirus	Parenteral or sexual transmission Insidious onset common; disease may be severe, with 1–2% mortality Associated with primary hepatocellular carcinoma, cirrhosis
Hepatitis C "Posttransfusion" hepatitis	**Enveloped** (+) ss RNA virus: Flavivirus	Parenteral or sexual transmission Acute disease is usually subclinical, but high rate of chronicity, with ~4% mortality Associated with primary hepatocellular carcinoma, cirrhosis
Hepatitis D "Delta" hepatitis	Defective **enveloped** RNA virus Requires hepatitis B as helper virus to replicate	Coinfection (both acquired at same time) is occasionally severe. Superinfection (patient already infected with B, then acquires B and D) has high mortality: cirrhosis, fulminant hepatitis

Chapter **21**

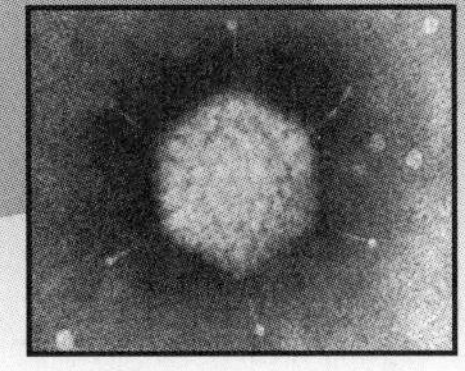

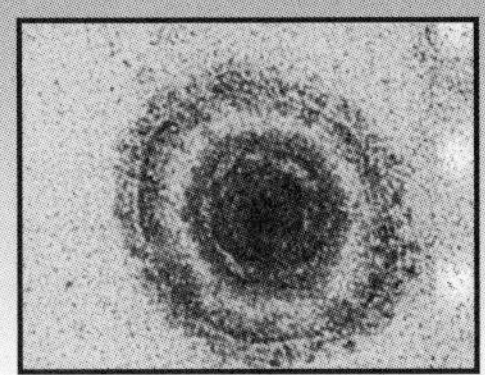

DNA Viruses

Naked: Parvoviruses, Papovaviruses, Adenoviruses
Enveloped: Hepadnaviruses, Herpesviruses, Poxviruses

I High-Yield Concepts

A. **ALL DNA VIRUSES, except the parvoviruses, are double-stranded** (ds).

B. **ALL DNA VIRUSES, except the poxviruses, duplicate their DNA in the nucleus** and are icosahedral.

C. **NAKED DNA VIRUSES are parvoviruses, adenoviruses, and papovaviruses.**

D. **TABLE 21-1** provides more details about both naked and enveloped DNA viruses.

II Replication

A. **DNA VIRUSES, except hepatitis B virus, duplicate their DNA by using it as a template to make more DNA.**

B. **HEPATITIS B is an enveloped ds DNA virus** that replicates through an RNA intermediate. (Ⓜ Remember the "peer pressure" from all the other hepatitis viruses—all RNA viruses replicating through RNA intermediates?) Hepatitis B's DNA polymerase has reverse transcriptase activity and makes the new DNA from the RNA intermediate.

TABLE 21-1 **DNA VIRUSES***

Virus Family	DNA Type	Virion-Associated Polymerase?	Envelope?	Shape	DNA Replicates in:	Major Viruses
Parvovirus	ssDNA	No	Naked	Icosahedral	Nucleus	B-19 Adeno-associated viruses
Papovavirus	dsDNA circular	No	Naked	Icosahedral	Nucleus	Papilloma Polyoma
Adenovirus	dsDNA linear	No	Naked	Icosahedral	Nucleus	Adenoviruses
Hepadnavirus	dsDNA circular (part ds)	Yes**	Enveloped	Icosahedral	Nucleus	Hepatitis B
Herpesvirus	dsDNA linear	No	Enveloped (nuclear)	Icosahedral	Nucleus	HSV Varicell-Zoster Epstein-Barr Cytomegalovirus
Poxvirus	dsDNA linear	Yes***	Enveloped (makes own)	Brick-shaped complex	Cytoplasm	Variola Vaccinia Molluscum contagiosum

* Mnemonic: Poor Poppie Adds Hop to Her Pox.

** Hepadnavirus replicates through an RNA intermediate which is used as the template to make more DNA. Therefore it carries its own DNA polymerase, which has RNA-dependent DNA polymerase activity.

*** A transcriptase, so that all the enzymes used for making and transcribing DNA are out in the cytoplasm where poxviruses replicate.

III Parvoviruses

A. **B19 is a naked, single-stranded (ss) DNA virus that replicates in erythrocytic precursors and causes fifth disease. Fifth disease (erythema infectiosum, a.k.a. slapped cheek fever)** most often infects adolescents, causing mild fever and **a recurring slapped cheek appearance with lacy rash on arms and then body.** It may cause **chronic anemia in immunocompromised** patients and **aplastic crises in sickle cell patients.** It can also cause **hydrops fetalis.**

B. **ADENO-ASSOCIATED VIRUSES are defective parvoviruses** not known to cause human disease that only replicate when coinfecting with adenoviruses. **Their ability to integrate into the human host genome has led to their use as vectors for the repair of defective genes.**

IV Papovaviruses

Papovaviruses are naked viruses with circular ds DNA. They include two groups (now classified as seperate families): the papilloma (wart) viruses and the polyomaviruses.

A. HUMAN PAPILLOMAVIRUSES cause warts and cervical carcinoma (serotypes 16 and 18). (Ⓜ Warts are circular, and you have to get "naked" to get the sexually transmitted human papilloma virus [HPV]. Also, the end result of some HPV infections is a positive *Pap* test.) HPVs are transmitted by direct contact.

1. **Plantar warts** are caused by HPV 1 and 4, which are both benign.
2. **HPV 6 and 11** are the most common causes of anogenital warts (condylomata acuminata, which are sexually transmitted) and laryngeal warts (usually seen in very young children and sometimes acquired at birth). They are problematic, regrowing after removal, but benign.
3. **Cervical intraepithelial carcinoma (and penile carcinoma) is most commonly associated with HPV 16 and 18. The early proteins of these oncogenic HPVs, E6 and E7, inactivate tumor suppressor functions of p53 and p110-Rb, respectively.**

B. POLYOMAVIRUSES BK AND JC are common but cause disease only in compromised patients. **BK virus causes kidney disease** and is often associated with kidney transplantation. **JC virus is associated with progressive multifocal leukoencephalopathy, a slow *conventional* viral disease of the immunocompromised and a problem in AIDS.**

V Adenoviruses

A. ADENOVIRUSES are naked, icosahedral ds DNA viruses with fibers projecting from the capsid's penton subunits. Adenoviruses have serotype numbers.

B. ADENOVIRUSES cause a variety of childhood and adult diseases, such as the following:

1. **Pharyngoconjunctivitis and keratoconjunctivitis.** Unlike bacterial pink eye, viral pink eye is not purulent; the conjunctivae are inflamed with a watery exudate.
2. **Acute respiratory diseases.** The most serious of these is interstitial pneumonitis in immunocompromised patients. The military has had to combat major problems with adenoviral respiratory disease in young recruits; **one unique, new, and effective vaccine consists of the virulent respiratory strains of adenovirus administered orally in enteric-coated capsules.**
3. **Adenoviruses 40 and 41** cause **gastroenteritis.**

VI Hepadnavirus

A. HEPATITIS B VIRUS is an enveloped DNA virus that replicates through an RNA intermediate. It carries its **own polymerase P, which has the reverse transcriptase, ribonuclease (RNase), and DNA-dependent DNA polymerase activity** in the mature virus particle. The genomic DNA is first transcribed into an RNA intermediate, which is encapsidated with the polymerase. In the capsid, the RNA is copied (using the reverse transcriptase) into the new genomic DNA, which is partially double stranded until it infects the next cell and can finish the replication.

B. TRANSMISSION. Hepatitis B is parenterally or sexually transmitted.

C. DISEASE MAY BE ACUTE OR CHRONIC. The **younger the age of infection, the higher the risk of chronicity. Chronicity** is indicated by the **presence of HBsAg for >6 months.** The correlation of various antigens and antibodies with the presence of virus and symptomology is shown in Figure 21-1. (Note: Patients with chronic infections are, on average, infectious for about 5 years and are unable to suppress the production of HBsAg.)

D. DIAGNOSIS is made by serology, testing for the presence of IgM to HBcAg, anti-HBsAg, and HBsAg. HBeAg's presence correlates well to the presence of virus. The presence of anti-HBeAg indicates a lower risk of transmission of the virus. (Please be careful in reading Hepatitis B serologies; anti-HBsAg can also be written as HBsAb, which can easily be misread as HBsAg.)

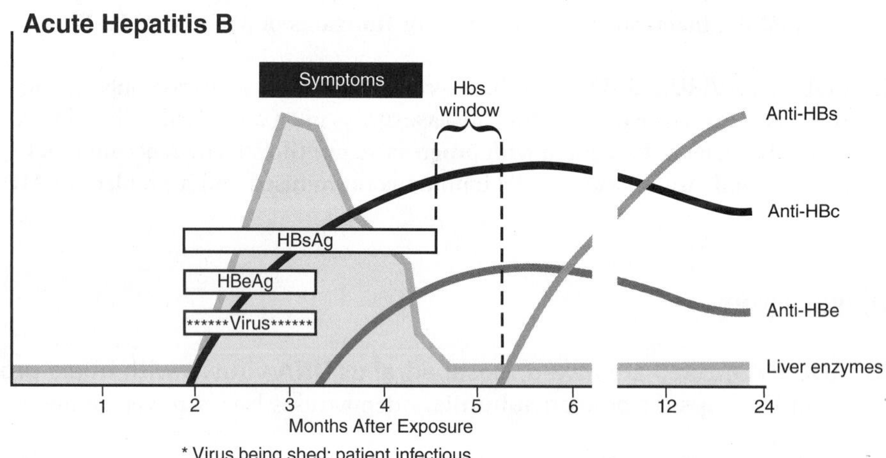

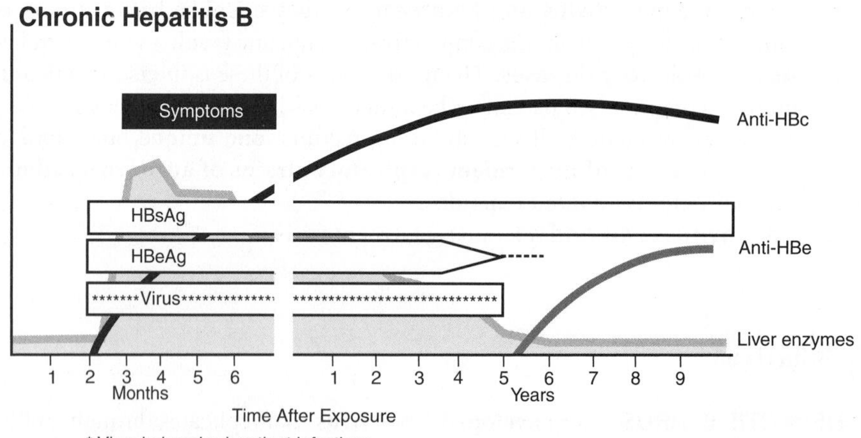

● **Figure 21-1** A comparison of virus production, antigens, antibodies, and symptoms of acute and chronic hepatitis B virus infections. These studies were important to allow serologic diagnosis. Note that the hepatitis B "e" antigen (HBeAg) correlates with presence of virus (infectivity) and that as soon as the virus disappears, the antibody to "e" rises. Also note that there is a time period (the window) when neither the "s" antigen nor the "s" antibody is detected; the antibody to "c" is used so as not to miss the diagnosis of hepatitis B virus.

 Herpesviruses

This family of large, **enveloped, icosahedral, ds DNA** viruses includes the herpes simplex viruses **(HSV)**, varicella-zoster virus **(VZV)**, Epstein-Barr virus **(EBV)**, and cytomegalovirus **(CMV).** Viruses of the herpes family are assembled in the nucleus of the infected cell; they bud first out of the host cell nuclear membrane, modified by viral glycoproteins, then discard that envelope and pass through the Golgi, picking up their final viral modified envelope. Herpesviruses may become latent: HSV and VZV both become latent in neurons, EBV in B cells, and CMV in lymphocytes.

A. HERPES SIMPLEX VIRUSES. HSV-1 and HSV-2 may cause either acute or latent infections.

1. **Latent infections.** The HSV DNA is present in nerve ganglia, and **only LAT proteins are expressed;** but the beta viral proteins, such as the **HSV thymidine kinase and DNA polymerase** (required for new DNA synthesis and for virus production), **are not made.** Thus, anti-HSV drugs like acyclovir, which require the HSV thymidine kinase to begin the conversion of the prodrug to the active drug to inhibit the HSV DNA polymerase, do not wipe out latent infections.
2. **Lesions.** Both **HSV-1** and **HSV-2** are noted for **recurrent vesicular cutaneous** lesions. A gross generalization is that HSV-1 usually infects above the waist and HSV-2 below the waist. HSV-1 is most noted for causing gingivostomatitis, keratoconjunctivitis, and encephalitis. HSV-2 is noted for genital herpes, neonatal infections, and meningitis.
3. **HSV-1.** Infections are often initially asymptomatic or manifest as gingivostomatitis (fever with painful oral vesicles or ulcers), usually in 3- to 5-year-olds. The virus may become **latent in the trigeminal nerve ganglia** and recur periodically as cold sores or fever blisters. Herpes may infect a finger or nail area, causing herpetic whitlow. **Herpetic ulcers of the cornea** or conjunctivae **may cause blindness.** HSV-1 is the most common cause of **sporadic encephalitis;** cerebrospinal fluid (CSF) may be normal, but the finding in CSF of slightly elevated lymphocytes and protein and **red cells is suggestive.** Polymerase chain reaction (PCR) on CSF is the diagnostic method of choice. Rapid diagnosis and administration of acyclovir have reduced mortality from >70% to <20%.
4. **HSV-2** is **sexually transmitted,** may be asymptomatic (both primary and recurrent), **localizes in sacral nerve ganglia (S3 and S4),** and may cause painful primary outbreaks and exacerbations with tingling prior to new outbreaks of **vesicular lesions,** which **ulcerate.** Regional lymphadenopathy occurs. Because of the risk of serious neonatal disease, delivery by cesarean section is indicated in the presence of active cervical, vaginal, or labial lesions. Neonatal herpes may be localized to skin, eyes, or mouth; it may be generalized, involving many organs, including the lungs, liver, and CNS; or it may be localized to CNS without other manifestations.

B. VARICELLA-ZOSTER VIRUS

1. **Primary infection: varicella or chickenpox.** This is usually a mild febrile disease (in normal children), characterized by asynchronous appearance of discrete, erythematous papular lesions on mucocutaneous surfaces. The **lesions vesiculate, develop into pustules, and crust over while new crops of lesions appear, producing the characteristic asynchronous rash.** The virus becomes **latent in neurons.** Incubation is 13 to 17 days. Children are considered contagious for 6 days after all lesions have dried. The attenuated ("live" but modified to reduce virulence) VZV vaccine reduces the incidence of both chickenpox and shingles.

2. **Secondary infection: herpes zoster or shingles.** This infection is characterized by the appearance of **clusters of vesicular lesions, usually along a single sensory dermatome, accompanied by nerve pain** that may last for months. Acyclovir is somewhat useful in both severe chickenpox and shingles.
3. **VZV-infected, immunocompromised patients.** Both chickenpox and shingles are **much more severe** in immunocompromised patients. Treatment is with varicella zoster immunoglobulin **(VZIG) *and* oral acyclovir.**

C. CYTOMEGALOVIRUS (CMV) is extremely common.

1. **Transmission. CMV may cross the placenta** (in either primary or reactivated infection), be **acquired during birth** or through **mother's milk**, or be transferred by **direct contact** with others, by **sexual** contact, or by **blood transfusion** or **organ transplant.**
2. **Replication.** When CMV replicates, it produces cells with typical **large, purple intranuclear inclusion bodies surrounded by a halo ("owl's eye")** plus smaller **basophilic cytoplasmic inclusions.**
3. **Disease.** Many CMV infections are **asymptomatic** (including later transplacental infections), but **in adults, they may be similar to mononucleosis (although heterophile-antibody negative).** CMV infections **in immunocompromised people are severe; clinical disease includes retinitis and interstitial pneumonitis.** In neonates infected very early, **cytomegalic inclusion disease** may develop (hepatosplenomegaly with thrombocytopenic purpura, pneumonitis, and CNS calcifications and microcephaly).

D. EPSTEIN-BARR VIRUS (EBV)

1. **EBV** is a Herpesviridae (herpes family) that **infects oral epithelial cells and B lymphocytes (the latter through CD21 = CR2).** EBV replicates in the epithelial cells and establishes a latent infection in B cells. Viral-specific antigens include the EBV nuclear antigen **(EBNA), which is found in all infected cells. In virus-producing cells, early antigens (EA) and viral capsid antigen (VCA) are also found.**
2. **In the United States, EBV infectious mononucleosis,** also known as the kissing disease, **is most commonly found in 15- to 25-year-olds.** It presents as severe **fatigue, pharyngitis/tonsillitis similar to strep throat, postcervical lymphadenopathy, hepatomegaly (with elevated alkaline phosphatase), and splenomegaly.** There is a dramatic mononucleosis with reactive **Downey type II cells (T cells)** and the formation of **heterophile antibodies (unique to EBV infections)** that react with animal red blood cell antigens (rather than with the virus). Heterophile antibody tests (e.g., Monospot) or tests for specific EBV antibodies to EA or VCA are available.
3. **EBV** is classified as a gamma herpes because of its **oncogenic potential.** It is **associated with Burkitt lymphoma,** which occurs in malarial areas of Africa; with a **nasopharyngeal carcinoma,** which occurs primarily in men from southern China; and with a lymphoproliferative disease in patients with primary or secondary immunodeficiencies. EBV is also associated with hairy cell leukoplakia of the tongue and interstitial pneumonitis in infants and lymphoma.

E. HUMAN HERPES VIRUSES 6 AND 8

1. In young children, human herpes virus 6 causes exanthem subitum (a.k.a. roseola), which is characterized by high fever for 3 to 7 days with a rash when the fever breaks (defervesces.)
2. Human herpes 8 (Kaposi sarcoma–associated herpes) infects B cells and is associated with Kaposi sarcoma. Seropositivity precedes sarcoma by about 3 years. The virus appears to be sexually transmitted.

Poxviruses

Poxviruses are very **large** and **unusual ds DNA viruses,** often described as having a **brick-shaped complex appearance.** Poxviruses **make all components, including their DNA, in the cytoplasm.** Thus, they **require a virion-associated transcriptase** in order to make the mRNA and the various viral enzymes in order to synthesize all their nucleic acids. (All human polymerases, ligases, and so on are located in the nucleus, so they cannot be used.) The poxviruses also direct the synthesis of their own envelopes, rather than simply modifying human cell membranes. Infection with poxviruses produces **cytoplasmic inclusion bodies.**

A. VARIOLA VIRUS caused **smallpox.** This virus has been extinct in the wild since 1977. (Ⓜ You should never see the **typical intracytoplasmic inclusion bodies**, called **Guarnieri bodies,** since they have "guone" away.)

B. VACCINIA VIRUS is the immunogen in the vaccine that led to the successful eradication of smallpox. Vaccinia is of uncertain origin.

C. MOLLUSCUM CONTAGIOSUM, another poxvirus, causes **benign, pink, pincushion-like skin tumors with nipplelike indentations.** Virus-infected cells have large **eosinophilic cytoplasmic inclusion bodies,** called ***molluscum bodies.*** This virus can be a problem in seriously compromised patients.

Chapter 22

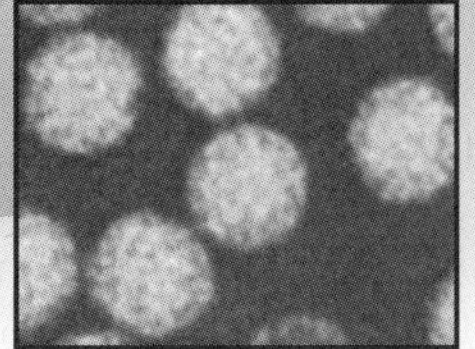

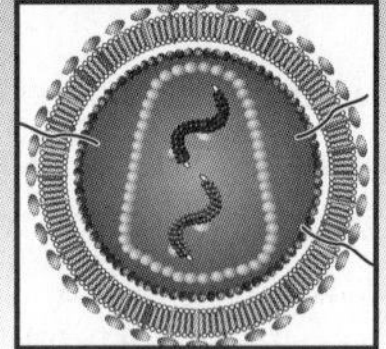

Positive (+) RNA Viruses

I High-Yield Concepts

A. **DEFINITION. Positive single-stranded RNA (abbreviated (+) RNA) viruses are those viruses whose genomic RNA either can serve directly as mRNA** (all positive RNA viruses except the retroviruses) **or has the same sequence as the mRNA** (retroviruses). All (+) RNA viruses (except the retroviruses) carry the genetic code for an RNA-dependent RNA polymerase. However, they do not carry the polymerase protein in the mature virion, because they can make it immediately in the host cell using the genomic RNA as their first mRNA.

B. **CHARACTERISTICS OF (+) RNA VIRUSES**
1. **Naked (+) RNA is infectious.**
2. **Positive RNA viruses (except retrovirus) replicate in the cytoplasm.**
3. **No (+) RNA virus is segmented.**
4. The **retroviruses are diploid** (two copies of the chromosome).
4. **Most RNA viruses are enveloped. The exception are the smallest—the caliciviruses and picornaviruses.**

C. **VIRAL FAMILIES.** See Table 22-1.

II Picornaviridae

Picornaviridae are a large group of **very small, naked, icosahedral (+) RNA viruses,** all of which replicate through a negative strand intermediate. Picornaviruses have very tightly fitting capsid protein subunits (capsomers), which are involved in cell binding and are **hard to inactivate with disinfectants or organic solvents.** (None of the required cell-binding proteins is held in place by lipid, as occurs in the enveloped viruses.) **Picornaviruses are divided into the Enterovirus genus (poliovirus, coxsackievirus, echovirus, enteroviruses 68 through 71, and hepatitis A, all of which survive stomach acid) and the rhinoviruses, which are acid-labile.** (Ⓜ A mnemonic for the picornaviruses is **PEECoRnA:** **P**olio, **E**ntero, **E**cho, **Co**xsackie, **R**hi**n**o, and hepatitis **A**.)

TABLE 22-1 POSITIVE-SENSE RNA VIRUSES

Virus Family*	RNA Structure	Virion-Associated Polymerase?	Envelope?	Shape	Multiplies in	Major Viruses
Picorna-virus	ss(+)RNA Linear nonsegmented	No polymerase	Naked	Icosahedral	Cytoplasm	Polio, Echo, Entero, Rhino, Coxsackievirus, hepatitis A
Calici-virus	ss(+)RNA Linear nonsegmented	No polymerase	Naked	Icosahedral		Norwalk agent, hepatitis E
Flavi-virus	ss(+)RNA Linear nonsegmented	No polymerase	Enveloped	Icosahedral	Cytoplasm	Yellow fever, Dengue, St. Louis encephalitis, hepatitis C
Toga-virus	ss(+)RNA Linear nonsegmented	No polymerase	Enveloped	Icosahedral	Cytoplasm	Rubella, west and east squire encephalitis, Venezuelan encephalitis
Corona-virus	ss(+)RNA Linear nonsegmented	No polymerase	Enveloped	Helical	Cytoplasm	Coronaviruses
Retro-virus	Diploid ss(+)RNA Linear nonsegmented	RNA-dep. DNA polymerase	Enveloped	Icosahedral or truncated conical	Nucleus	HIV, human T cell lymphotrophic virus, sarcoma

*Mnemonic: Pico Calls Flavio To Come Right Away.

A. **POLIOVIRUS**
   1. **Poliovirus transmission is fecal-oral** (contaminated swimming holes).
   2. **Serotypes.** There are **three stable polioviruses (1, 2, and 3),** which are close to being eradicated through the use of the **oral (attenuated) Sabin or the killed (Salk) vaccines.**
   3. **Disease.** The virus infects **oral and gastrointestinal (GI) epithelium;** >90% of the infections are asymptomatic. When the virus infects the anterior horn cells of the spinal cord (1 in 250 infections), poliovirus may cause a paralytic disease involving the cranial and respiratory nerves. Because of cases of vaccine-associated paralytic poliomyelitis (i.e., caused by the vaccine strain), the United States now uses the inactivated (Salk) vaccine.

B. **OTHER ENTEROVIRUSES (coxsackie A and B, enteroviruses 68–71, and echoviruses)** are common and significant **causative agents** of the following:
   1. **Upper respiratory infections** (URIs), such as **colds, herpangina, and stomatitis.** Herpangina is characterized by gray-white papulovesicles on anterior tonsillar pillars and tonsils, the soft palate, and the uvula.
   2. **Rashes and vesicles,** such as **hand-foot-and-mouth disease,** which is a vesicular disease.
   3. **Central nervous system** (CNS) infections—**most often meningitis;** less commonly encephalitis. **Viral meningitis (usually benign) presents with fever, headache, stiff neck, and malaise. Cerebrospinal fluid (CSF) cell counts are 0 to 500 cells per mm^3 (initially polymorphonuclear neutrophil leukocytes [PMNs] predominate, then lymphocytes); glucose is generally near normal, and protein is normal or only slightly elevated.**
   4. **GI disorders**
   5. **Hemorrhagic conjunctivitis**
   6. **Myopericarditis—caused primarily by Coxsackievirus B and characterized by chest pains, fever, fatigue, and arrhythmias**

C. **HEPATITIS A is a picornavirus.** (For a review of hepatitis viruses, see Table 20-2.)

D. **RHINOVIRUSES** are the major causative agents of the **common cold.** They peak in the **summer and early fall.** (Ⓜ Think of the rhinoceros on the Serengeti. It is hot!) They are not resistant to stomach acid.

III Caliciviridae

Caliciviridae are naked icosahedral (+) RNA viruses and are slightly larger than the picornaviruses. Caliciviruses cause gastroenteritis and hepatitis E.

A. **NORWALK AGENT GASTROENTERITIS** (sometimes called Norovirus) is a noninflammatory diarrhea found in **older children and adults.** It is often a fecal-oral spread in food or by contaminated shellfish, but in developing countries, it may be waterborne. There is usually a 1- to 2-day incubation and **1 day of illness, with vomiting, diarrhea, and a low-grade fever.**

B. **HEPATITIS E** (see Table 20-2) has a **high fatality rate in pregnant women.**

IV Flaviviridae

Flaviviridae are enveloped, icosahedral (+) RNA viruses. (Ⓜ *Flavi* means "yellow," and this family has viruses that "turn you positively yellow.")

A. **HEPATITIS C ("posttransfusion") often is initially asymptomatic, but it has a high rate of chronicity** (see Table 20-2). Blood donors are now screened for hepatitis C and therapeutic intravenous immunoglobulins are treated to inactivate hepatitis C. Thus, it is now **spread mainly by needle sharing.** Diagnosis is by screening enzyme immunoassay with a recombinant immunoblot (similar to HIV diagnosis), as well as polymerase chain reaction (PCR) for RNA. There are at least **six genotypes. Treatment is with interferon alpha and ribavirin. Genotype 1 does not respond well.**

B. **YELLOW FEVER VIRUS** (tropical South America and Africa) **is spread by mosquitoes** (arthropod-borne virus = arbovirus). It causes febrile disease of varying severity, which often includes hepatitis and which may progress to hemorrhagic fever. The vaccine is attenuated.

C. **DENGUE VIRUS** is another arbovirus belonging to the Flaviviridae. Dengue (breakbone disease) is spread by ***Aedes* mosquitoes in the tropics,** including the Gulf Coast of the United States. Incidence of the disease is increasing. Clinically, **Dengue has a rapid onset, with fever, severe myalgias, arthralgias, headache, and rash.** There are four serotypes. **Dengue hemorrhagic shock syndrome (DHSS) is a serious complication seen in young babies with maternal antibodies or in any individual with antibody from a previous infection who is infected with a second dengue serotype.** Antibody to one serotype appears to increase the ability of a second serotype to infect cells. Thus, DHSS occurs in infants and then decreases, gradually increasing with age and exposure.

D. **ST. LOUIS ENCEPHALITIS VIRUS** is spread by the *Culex* mosquito and is endemic in Canada, the United States, the Caribbean, and South America. The disease is **most severe in the elderly.**

E. **WEST NILE VIRUS** spread to the western hemisphere in 1999 and is now a major **mosquito-borne infection of birds, horses, and humans.** Although thousands of humans have been infected, **deaths have occurred mainly in debilitated elderly persons.** In addition to the encephalitis, there may be muscle weakness and flaccid paralysis.

V Togaviridae

Togaviridae are enveloped, icosahedral (+) RNA viruses. (Ⓜ Think of a handsome [a positive thing = (+) RNA] Roman man with spots [rubella] who is wearing a toga [envelope] and riding his horse [dead-end host for encephalitis viruses], with birds [the reservoir for the equine encephalitis viruses] and mosquitoes [the vector] flying around him.)

A. **ALPHAVIRUSES** (the mosquito-borne Toga viruses) include the **Western, Eastern, and Venezuelan equine encephalitis (EE) viruses. Horses are dead-end hosts; wild birds are the normal hosts. Mosquitoes transfer the disease to humans.**

B. **RUBELLA VIRUS** causes rubella, or the "German measles," which is characterized by a **discrete red maculopapular rash, lymphadenopathy, and mild fever.** The virus **can cross the placenta and cause serious birth defects,** such as cataracts and retinopathy, patent ductus arteriosus, sensorineural deafness, and retardation. This disease is completely **vaccine preventable.** The rubella component of the measles-mumps-rubella vaccine (MMR) is a **single attenuated strain of rubella virus.** It is recommended that barrier protection be used to prevent pregnancy for 16 weeks after vaccination.

VI Coronaviridae

Coronaviridae are larger, enveloped, helical (+) RNA viruses that are divided into antigenic groups. The envelope has **prominent surface glycoproteins (hemagglutinins), and the virus resembles a crown** (hence "corona").

A. **COMMON COLDS.** Coronaviruses are the second most common causative agents of the **common cold** and appear mainly in **winter and spring.** (Ⓜ Coronavirus infections peak in cold weather, so think of a cold Corona beer.) Coronavirus infections tend to have more nasal discharge as compared with rhinovirus infections, which tend to have more cough.

B. **SEVERE ACUTE RESPIRATORY SYNDROME (SARS)** first appeared in late 2002 in China and has since spread worldwide. The SARS-coronavirus causes severe pneumonia.

Retroviridae

Retroviridae are diploid, enveloped (+) RNA viruses that replicate through a DNA intermediate. Within the virion envelope, the virus carries the **reverse transcriptase, protease,** and **integrase proteins,** as well as the **nucleocapsid and capsid proteins.** Human T-cell leukemia virus 1 (HTLV-1) and HTLV-2 are oncoviruses (retroviruses that have lost some genes and that carry oncogenes). They infect cells but cannot make virus. However, they can cause transformation of cells and are associated with human leukemias. Human immunodeficiency virus types 1 and 2 (HIV-1 and HIV-2) are lentiviruses (slow viruses) that do not carry extra oncogenes. HIV-1 is currently the medically most important retrovirus and is discussed below.

A. GENES* AND GENE PRODUCTS OF IMPORTANCE IN HIV INFECTION

1. The *gag* gene (group specific antigens) codes for the structural proteins p24 (CA = capsid); p7 and p9 (NC = nucleocapsid proteins); p17 (MA = matrix protein), which is membrane stabilizing; and the protease (PR), which is critical in post-translational processing of the other gag proteins so they can be used as structural proteins.
2. The *env* gene codes for surface (SU) glycoprotein gp120 (which binds CD4) and transmembrane (TM) protein gp41 (which early on can bind to one of the chemokine receptors, CXCR4, and infect predominantly macrophages or later can bind to CCR5, which increases the T cell tropism). TM also plays a role as a fusion protein, allowing infected cells to fuse with uninfected cells to open up new cellular factories without leaving the cell.
3. The *pol* gene codes for the integrase and reverse transcriptase (which also carries the ribonuclease H activity).

*Gene abbreviations are italicized, as is standard.

4. Long terminal repeats (LTRs) are critical to integration and regulate transcription via active enhancer promoter regions.
5. Additional regulatory genes, such as *tat* (transactivator of transcription) and *rev*, are positive regulators of viral production and play critical roles in disease progression. Infected cells (if not treated with effective antiretroviral drugs) continue to produce low levels of HIV. Coinfection with something like *Mycobacterium tuberculosis* causes increased levels of virus production, because T cell activation increases the ability of the *tat* protein to produce whole mRNA The *nef* protein down-regulates cell markers (both CD4 and MHC I) on cells.

B. EVENTS OF HIV INFECTION IN AN UNTREATED PERSON

1. **HIV SU** (gp 120) binds to CD4 molecules and TM binds to one of the chemokine receptors and permits entry. Membrane fusion, involving chemokine receptors binding TM to allow nucleocapsid entry, is followed by partial uncoating of the genomic RNA. The reverse transcriptase makes the ds DNA "copy," which migrates into the cell nucleus. Integrase activity inserts the viral DNA into the human chromosome.
2. Following transcription and production of early proteins, HIV proteins regulate transcription, RNA cleavage and transport of RNA to the ribosomes, and protein cleavage, all of which play roles in controlling viral production. The cleaved mRNAs are transported to the cytoplasm, along with the intact genomic RNAs. After the proteins are made, SU and TM are inserted into the host cytoplasmic membrane. The virus is assembled and matures through the membrane to pick up the envelope.
3. Clinically, there is an initial acute mononucleosis-like disease, with fever, fatigue, malaise, arthralgias, and rash, with high titers of circulating virus. This is followed by a relatively asymptomatic period with low viral titers in the blood once the immune system is primed. Infections with other organisms that activate infected T cells (e.g., hepatitis B or tuberculosis) increase production of HIV. When CD4+ cell counts (and antibodies to HIV) decline, viral production goes up.

C. OPPORTUNISTIC INFECTIONS. Most serious infections become major problems at a CD4+ level <200 mm^3. Major infective organisms are listed below. Routine prophylaxis is used to prevent *Pneumocystis carinii (P. jiroveci), Mycobacterium avium-intracellulare,* and *Toxoplasma* infections.

1. Fungal infections include *Candida* (oral thrush → gastritis → septicemia), *Cryptococcus meningitis* (encapsulated yeast in CSF), *Histoplasma* or *Coccidioides* (depending on exposure, sometimes years earlier), and *Pneumocystis* infections. Pneumocystis prophylaxis is recommended when patients' CD4+ counts drop <200.
2. Bacterial infections include disseminating pulmonary infections with *M. tuberculosis* and *M. avium-intracellulare,* as well as gastroenteritis caused by *Salmonella, Shigella,* and *Campylobacter*. Prophylaxis for nontuberculous mycobacterial infections is recommended for those with CD4+ count <100.
3. Viral infections include herpes simplex virus 1 and 2 infections and cytomegalovirus (CMV) chorioretinitis and pneumonia.
4. Protozoan infections with *Cryptosporidium, Giardia, Toxoplasma,* and *Entamoeba* are common.

Chapter 23

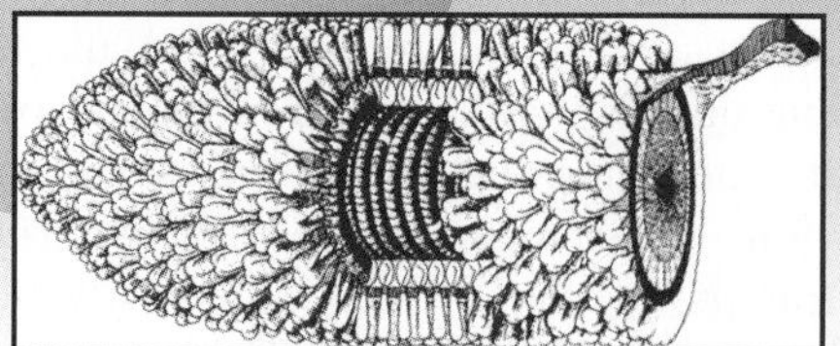

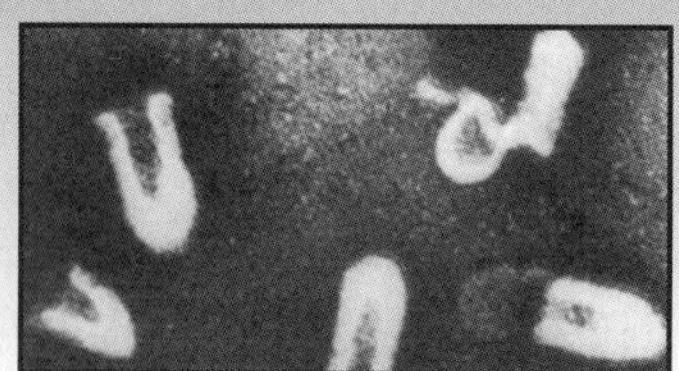

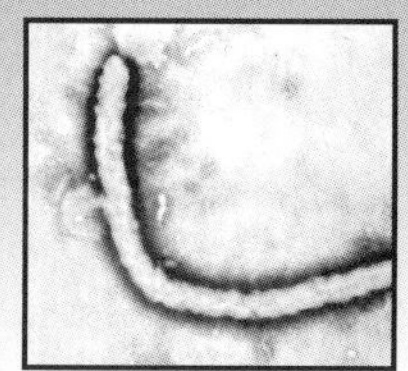

Negative (–) Single-Stranded RNA Viruses

I High-Yield Concepts

A. **NEGATIVE-SENSE RNA, abbreviated (–) RNA, is the homolog of mRNA.**

B. **ALL (–) RNA VIRUSES carry a virion-associated, RNA-dependent RNA polymerase** (the protein, not just the gene) that directs production of the (+) RNA homolog, which is used both as mRNA and as a template to make more negative copies for progeny virus.

C. **NAKED (–) RNA (i.e., without the RNA-dependent RNA polymerase protein) is not infectious.** Humans do not have RNA-dependent RNA polymerase available to use.

D. **ALL (–) RNA VIRUSES are helical and enveloped. No (–) RNA viruses are naked or icosahedral.**

E. **THREE FAMILIES OF (–) SS RNA VIRUSES are segmented: Orthomyxoviridae, Bunyaviridae, and Arenaviridae.** The double-stranded (ds) RNA viruses (Reoviridae) are also segmented. (Ⓜ ROBA [like robots, made of pieces]: Reoviridae, Orthomyxoviridae, Bunyaviridae, Arenaviridae.) The (–) RNA viral families are shown in Table 23-1.

II Replication

A. **VIRAL SURFACE GROUPS held by the envelope determine how these viruses enter the host cells.** Viruses with fusion (F) proteins fuse their viral envelope with the host cell membrane to enter cells. The remaining enveloped viruses enter by a pinocytotic process triggered by virus envelope glycoproteins (such as hemagglutinins) binding to host cell receptors.

B. **ENVELOPE GLYCOPROTEINS also determine whether the virus is hemagglutinating or nonhemagglutinating.** In terms of host antibody response, if the hemagglutinin (H) and neuraminidase (N) are separate, each stimulates a separate antibody.

TABLE 23-1 **NEGATIVE-SENSE RNA VIRUSES**

Virus*	RNA Structure	Virion-Associated Polymerase?	Envelope?	Shape	Multiplies in	Major Viruses
Paramyxovirus	Linear nonsegmented ss(–)RNA	Yes†	Yes	Helical spikes: HN and F glycoprotein	Cytoplasm	Measles, mumps, respiratory syncytial, parainfluenza
Rhabdovirus	Linear nonsegmented ss(–)RNA	Yes	Yes	Bullet-shaped helical	Cytoplasm	Rabies, vesicular stomatitis
Filovirus	Linear nonsegmented ss (–)RNA	Yes	Yes	Helical		Marburg, Ebola
Orthomyxovirus	Linear 8 segments ss(–)RNA	Yes	Yes	Helical	Cytoplasm and Nucleus	Influenza A and B
Bunyavirus	Linear 3 segments: 2 ss(–) RNA 1 ambisense	Yes	Yes	Helical	Cytoplasm	California encephalitis, La Crosse, Hanta, Si Nombre
Arenavirus	Circular 2 segments: 1 ss(–) RNA 1 ambisense	Yes	Yes	Helical	Cytoplasm	Lymphocytic choriomeningitis, Lassa fever

* Mnemonic: **Para Ra**bbits **Fi**ght **O**ver **Bun**nies' **Ar**ea (fighting is a negative thing)
or **P**ara **R**abbits **F**ill **O**rthodox **B**unyan **A**rena

† In all cases RNA-dependent RNA polymerase.

(Note: This is important in vaccine or protection questions [e.g., influenza A/H1N1]. The presence of a number between an H and an N designation should remind you that these are separate antigens.)

III Paramyxovirus

Paromyxoviruses are large, enveloped, helical (–) RNA viruses. They have a single dominant serotype, except for the parainfluenza viruses, which have four. All are transmitted via respiratory secretion, directly or by fomites.

A. MEASLES VIRUS (RUBEOLA)

1. **Rubeola has H, F (no N), and a virion-associated polymerase.**
2. **Measles** presents with **fever, cough, coryza, conjunctivitis, Koplik spots** (gray-white with red base on oral mucosa), **and an erythematous maculopapular rash (resulting from the action of cytotoxic T cells on infected cells in the microcapillaries). The rash starts on the face and moves down, becoming quite confluent.**
3. **Measles complications** are more common in younger, malnourished, or immunocompromised children. (Vitamin A treatment increases survival in malnourished children.) Complications **include otitis media, bronchopneumonia with giant cells** (Warthin-Finkeldey cells), and diarrhea. In a small percentage, persistent infection leads to **subacute sclerosing panencephalitis** years after infection.
4. **The measles vaccine is an attenuated strain.**

B. MUMPS VIRUS has a virion-associated polymerase, an F protein, and a combined HN protein.

Clinical disease includes **salivary gland enlargement in most cases, meningeal signs, and orchitis (postpuberty and rarely causing sterility).** The **mumps vaccine is attenuated.** Mumps encephalitis has become rare with the vaccine.

C. PARAINFLUENZA VIRUS, like mumps, has combined HN, F, and a virion-associated polymerase.

It causes **croup (laryngotracheobronchitis)** in infants, **upper respiratory infections (URIs), pneumonia,** and **bronchiolitis.** There are four serotypes and no vaccine.

D. RESPIRATORY SYNCYTIAL VIRUS (RSV) is an enveloped negative (–) RNA virus with only an F protein.

RSV is the major cause of bronchiolitis and pneumonia in infants and young children; it also causes colds with bronchitis in older children, adolescents, and adults.

IV Rhabdoviridae

A. RABIES VIRUS is an enveloped (–) RNA virus with a helical nucleocapsid and overall bullet shape.

1. **Epidemiology.** *U.S. reservoirs* **of rabies are skunks (in the west), raccoons or foxes (east), coyotes or dogs (Texas–Mexico border), and bats (throughout the country). Rabies** *vectors:* **Rabies is usually transmitted to humans by bat contact or dog bite** (an unvaccinated dog having acquired it from a reservoir). Rodents and rabbits rarely carry rabies. Over the past 20 years, about 25% of U.S.

human rabies deaths resulted from bites (89% of those dog bites). For 75% of the deaths, no bite was known; more than half were found to be **insectivorous bat** strains. Aerosals of bat rabies in caves are a risk for spelunkers (cave explorers.)

2. **Pathogenesis. Rabies virus may grow in muscle from days to months** before infecting the peripheral nervous system, where it moves by **axonal transport to the central nervous system (CNS),** replicates in the gray matter, and descends to infect eyes, salivary glands, skin, and other organs.
3. **Postexposure treatment.** Risk assessment in conjunction with Public Health Officers for all bites, contacts with bats, or potential exposures is critical. This is an enveloped virus, so cleaning any wound with detergent or soap and water, followed by alcohol and then povidone-iodine (virucidal), is important. **Hyperimmune serum is injected in and around the wound site, and administration of the human diploid vaccine (inactivated; five doses) at a different site may prevent disease.** This vaccine is used for preexposure in veterinarians, speleunkers, animal control people, and so forth.
4. **Negri bodies are the eosinophilic intracytoplasmic inclusion bodies** seen in tissues. **Fluorescent antibody staining** is used to examine brains of potentially infected animals as well as skin biopsies taken from the nape of the neck for suspected human cases.

B. VESICULAR STOMATITIS VIRUS is an arbovirus that causes foot-and-mouth disease; it is uncommon in humans.

V Filoviridae

Filoviridae viruses (a.k.a. "thread" viruses) are **long, helical, enveloped (–) RNA viruses.** Both **Ebola** and **Marburg** viruses are causative agents of **hemorrhagic fever, a disease that starts with influenzalike symptoms, then vomiting and diarrhea, and often ends with severe bleeding, shock, and death.**

VI Orthomyxoviridae (Influenza A and B Viruses)

A. INFLUENZA A AND B VIRUSES are enveloped, segmented (–) RNA viruses with eight helical segments. Separate viral-coded H and N glycoproteins stud the envelope surface, playing the usual roles in cell binding and release. **The hemagglutinin appears to induce the most protective antibodies.**

B. GENETIC DRIFT. Because the **RNA-dependent RNA polymerase makes errors without repairing them (no editing function), the resulting minor surface antigen changes (referred to as genetic drift) mandate an annual vaccine program** that updates the immune system on the newest strains of both influenza A and B viruses. (See Chapter 25, Section IA, for more information on genetic drift.)

C. GENETIC REASSORTMENT. Periodically, there is a dramatic antigenic change in the surface glycoproteins of influenza A viruses (called *antigenic shift* or *genetic reassortment*) that leads to pandemics (often serious worldwide outbreaks of influenza A). (Note: There are eight different segments of RNA, or "chromosomes," in each virus.) **Antigenic shift requires that an individual cell is coinfected with two different**

strains of influenza A such as avian and human. The progeny virus is assembled with some "chromosomes" of both strains (see Figure 25-1). This disease often starts in rural Asia, where people live close to their pigs and chickens, and results in the production of a dramatically new strain for which humans do not have any pre-existing partial immunity. This leads to a **pandemic.**

VII Bunyaviridae

Bunyaviridae are **segmented, enveloped, helical (–) RNA viruses** with **two chromosomes of ss (–) RNA** and **one ambisense** chromosome. (Some regions of ambisense RNA function like (–) RNA and others like (+) RNA.)

A. CALIFORNIA ENCEPHALITIS VIRUS (spread by ***Aedes*** spp. mosquitoes) causes subclinical infections or occasionally **nonfatal clinical encephalitis in North American school-aged** children in the summer and fall. **Small mammals are the reservoir hosts.**

B. HANTAVIRUS (SIN NOMBRE HANTAVIRUS) causes an infection with an **influenza-like presentation, leading to acute respiratory failure (hantavirus pulmonary syndrome).** It is transmitted in **deer mice feces and urine.**

VIII Arenaviridae

Arenaviridae are **enveloped RNA viruses** with **two helical segments: one (–) RNA** and **one ambisense RNA.**

A. LYMPHOCYTIC CHORIOMENINGITIS (LCM) virus causes an **influenzalike febrile disease with meningitis.** Unlike most viral meningitis, **LCM may be fatal.**

B. OTHER ARENAVIRUSES cause Lassa fever and the South American hemorrhagic fevers (Argentinean and Bolivian). These febrile diseases are complicated by hemorrhage.

Chapter 24

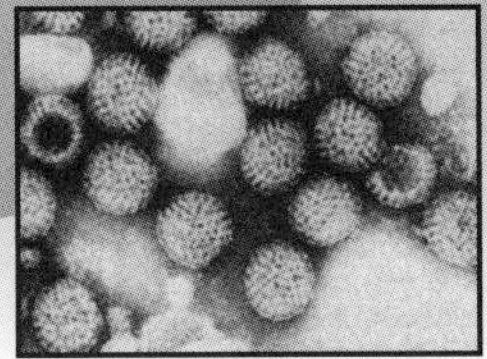

Double-Stranded RNA Viruses: Family Reoviridae

I Major Concepts

Viruses of the **family Reoviridae are naked** and have both **double-stranded (ds) RNA** and **double-shelled icosahedral capsids.** There are 10 to 12 segments (chromosomes) of ds RNA, depending on the virus. There are three viruses of importance to humans: Rotavirus, Orbivirus, and Reovirus.

II Rotavirus

Rotaviruses have **11 segments of ds RNA** and are the **major cause of infantile diarrhea worldwide. Vomiting and fever** (most commonly 37.9°C to 39°C) **usually precede diarrhea.** In healthy infants, **illness lasts 3 to 9 days.** The major serotype (A) causes diarrhea and dehydration in infants <2 years, mild diarrhea in older kids, and severe diarrhea, leading to death, in malnourished individuals. This **noninflammatory diarrhea is caused by protein nsP4, a viral enterotoxin.**

III Orbivirus

The **Colorado tick fever virus**, an Orbivirus species transmitted by *Dermacentor andersoni,* is one causative agent of **viral encephalitis.**

IV Reovirus

Exposure to the reoviruses (sometimes called the **orthoreoviruses**) **without overt disease** appears to be common in humans. These viruses may also cause a febrile disease.

Chapter 25

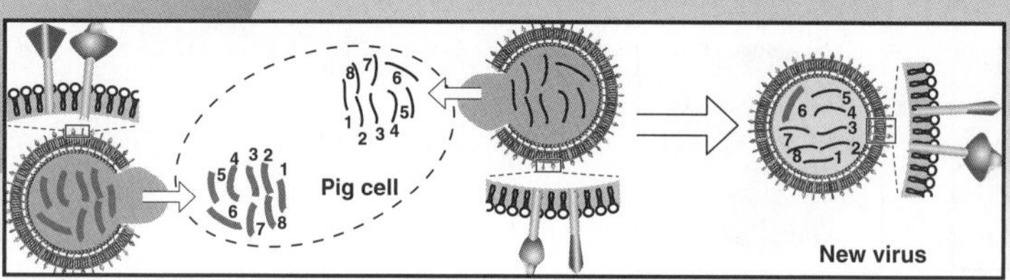

Viral Genetics

I Mutation: Genetic Drift and Defective Interfering Particles

Mutations are caused by genetic errors. Mutational change can help viruses evade the immune system; they can also limit the duration of a viral infection. Most **viral-coded polymerases do not recognize and fix errors.** As a result, mutation rates are higher in viruses that make their own polymerases than in viruses that use human host cell polymerases. Viruses that make their own polymerases include all the RNA viruses, poxviruses, herpes simplex virus, varicella-zoster virus, and adenoviruses. A **point mutation** is one type of genetic error. Other errors include larger deletions or additions made when the polymerases "fall off" the template and reattach at a different site on the template; this leads to **shorter or longer defective pieces of nucleic acid.** Medically important genetic errors include the following:

A. GENETIC DRIFT. Point mutations lead to **minor changes in antigenicity called *antigenic drift.***

1. **Influenza viruses.** Antigenic drift in the influenza virus hemagglutinins and neuraminidases necessitates new influenza vaccines every year.
2. **Human immunodeficiency virus (HIV).** The antigenic drift in HIV's **gp120** (the major surface antigen) leads to multiple variants within one person dying of AIDS. Antigenic drift helps HIV evade the immune system and has slowed vaccine development.

B. DEFECTIVE INTERFERING PARTICLES. Virus production continues as long as there is one functional copy of every viral gene in the host cell. As replication proceeds, more defective nucleic acids are produced. Polymerases bind to both the normal and the defective nucleic acid pieces. The shorter defective nucleic acids replicate faster, leading to increased production of viral particles with defective nucleic acid, known as **defective interfering (DI) particles.** As infection progresses, fewer normal virions and more DI particles are produced, helping to limit the viral infection.

II Viral Complementation

Viral complementation (coinfection of one cell with two viral genomes leading to virus production) is a genetic laboratory technique but also a real-life phenomenon. Possible types include the following:

A. COINFECTION WITH BOTH DI PARTICLES AND NORMAL VIRUS. This was described in Section IB. As viral production proceeds, fewer normal virus and more DI particles are produced, until there is no longer a functional copy of each gene in the cell.

B. COINFECTION WITH TWO MUTANT (AND NO NORMAL) STRAINS OF THE SAME VIRUS

1. **Continuation of virus production (complementation)** indicates that each mutant was defective in a different gene, so there was one functional copy of each gene in the cell to allow production of all proteins. (Unless recombination occurs, all of the virions will still have mutant nucleic acid.) Complementation is like having two nonfunctional 1961 Corvettes in your garage. If each is defective in a different part, you can pirate parts and drive out in a functional car.
2. **Lack of viral production** indicates each of the two strains was defective in the same gene, so one protein is missing.

C. COINFECTION WITH DEFECTIVE VIRUS AND HELPER VIRUS. **Hepatitis D,** a defective RNA virus, can only replicate in a cell coinfected with its **"helper" virus hepatitis B** (a DNA virus). Hepatitis B replication provides the surface protein that hepatitis D needs for infectivity.

Gene Reassortment (Genetic Shift)

Genetic shift occurs with some **segmented viruses,** such influenza A viruses with multiple hosts. There are many influenza A avian, human, and animal strains. Where farmers live in close contact with their animals, **coinfection of pig cells with both a human and an avian strain may occur. If a virus is assembled with the neuraminidase gene segment of one virus and the hemagglutinin of the other, this gene reassortment creates a dramatically different strain,** which can lead to worldwide outbreaks of very severe disease (pandemics), such as the influenza epidemic of 1918. See Figure 25.1.

Latency

Latent viruses exist as **stable nucleic acid** in the host cell, either **free in a plasmidlike state or integrated into the host cell genome (provirus).** Continued production of repressor proteins prevents the replication of virus.

A. LYSOGENY. Latency of bacteriophage in bacterial cells is called **lysogeny** (see Chapter 5).

B. LYSOGENIC CONVERSION occurs when, in addition to the repressor protein, other phage-borne genes are expressed, **making the bacterium virulent.** Medically important examples are modification of *Salmonella* **O antigens** and production of **Botulinum toxin** (here, the phage may actually be defective), the **erythrogenic toxins of *Streptococcus pyogenes*** (SPE-A, B, or C), and **Diphtheria toxin.** (Ⓜ **OBED,** pronounced o-BEED: These bacteria are a little bit "pregnant" with phage DNA.)

C. ENTEROHEMORRHAGIC *ESCHERICHIA COLI* SHIGALIKE TOXIN PRODUCTION. Enterohemorrhagic *E. coli* (EHEC) strains are lysogenized with a phage, but toxin production occurs when the phage exits lysogeny and starts lytic production of the virus.

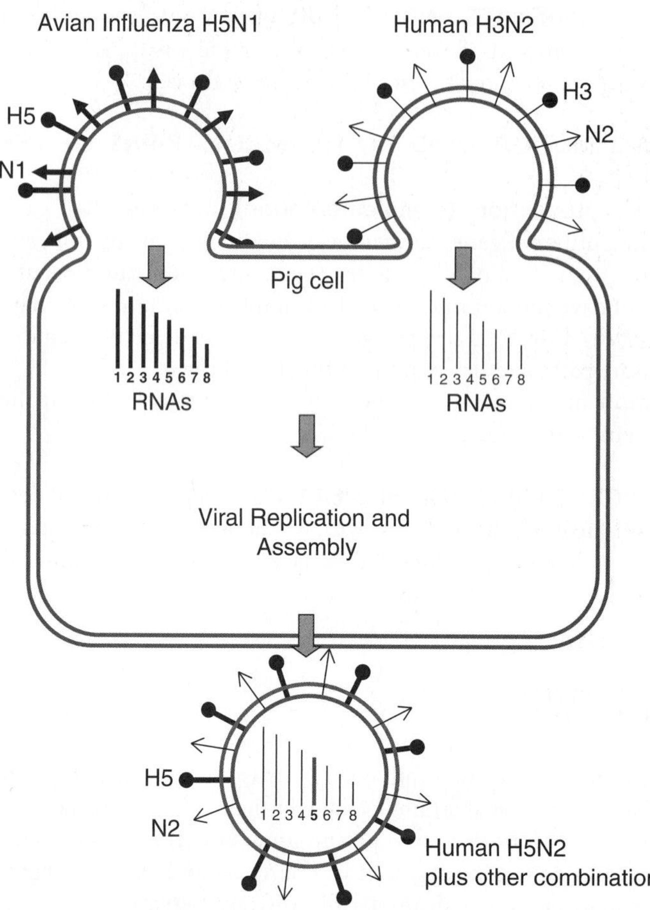

Genetic Reassortment = Genetic shift

1. Two influenza A viruses from different animals co-infect one cell. Each of the 8 RNAs carries code for a gene product.

2. NOT SHOWN: Each virus replicates all RNAs and all proteins.

3. (Remember two viruses have replicated in one cell.) In the viral assembly, the viral RNAs package randomly (genetic reassortment) which may produce a new virus which now can infect humans (e.g., H5N2) and to which there is no pre-existing immunity.

● **Figure 25-1** Major antigenic changes called genetic shift or gene reassortment, occurr when a single cell is coinfected with two different strains of the same segmented virus.

The EHEC toxin is produced with the phage components, and these phage can infect other *E. coli.* Antibiotics, such as fluoroquinolones and trimethoprim, which activate the SOS repair system, increase the production of the toxin and thus increase the infected person's risk of hemolytic uremic syndrome (HUS).

D. RECURRING INFECTIONS. Latency in human viruses allows for persistent or recurring infections. Latent human viruses include Herpesviridae and some Adenoviridae. HIV, although inserted, is not truly latent, because it is continuously expressed at a low level, unless the patient is being treated with inhibitors.

E. ONCOGENESIS. Latent viruses may play a role in human cancers through the following:

1. Production of early oncogenic proteins (e.g., human papilloma virus)
2. DNA insertion, which results in mutagenesis of a cellular regulatory gene and leads to a loss of growth control
3. Up-regulation, by insertion of a viral transcriptional activator, such as HIV's LTR, near human genes that regulate cell growth

Phenotypic Masking and Mixing

When a cell is coinfected with two normal (not defective) and often related viruses, two interesting phenomena may occur. To visualize this, think of one as a red virus with red capsomers and red DNA and the other as a green virus with green capsomers and green DNA.

A. PHENOTYPIC MIXING. If the geometry of the heads is very similar, virus may be produced with capsids made up of a mixture of green and red capsomers. Only one kind of DNA (green or red) would be present in each capsid.

B. PHENOTYPIC MASKING occurs when an entirely red capsid forms around a green viral genome; the virus is called a *composite virus*. The **only way to produce composite viruses** (capsid of one and nucleic acid of another) is for the cell to be coinfected with the genomes of both viruses. (Note: Composite viruses are used on the USMLE Step 1 exam to test whether you know that the nucleic acid directs the production of progeny viruses and that the host cell range is determined by the outer glycoproteins or proteins.)

Transfection

In the lab, naked viral DNA or positive RNA can be used to infect human cells in culture and initiate viral infection. This is called **transfection.** Naked negative RNA (without the polymerase protein) is noninfectious.

Viruses as Vectors

Viruses may circulate in the body and bind to specific cell receptors, triggering viral uptake and subsequent release of the nucleic acid inside specific cell types. These viral activities are useful to genetic engineers, who create **viral vectors** to deliver genes to repair specific cell types. The needed gene is inserted into the viral vector, which is modified so the viral vector is unable to replicate. Delivery of genes via viral vectors is not without problems, such as the production of viral-neutralizing antibodies, which might lead to an adverse reaction. In addition, defective viruses present in the human may complement the vector, allowing replication and spread of the modified nucleic acid. However, viral vectors do show promise in **gene repair** and in **delivery of drug-producing genes** to specific cells, such as brain tumor cells.

Part **IV**

Fungi

Chapter 26

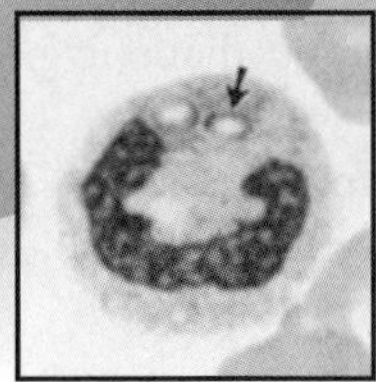

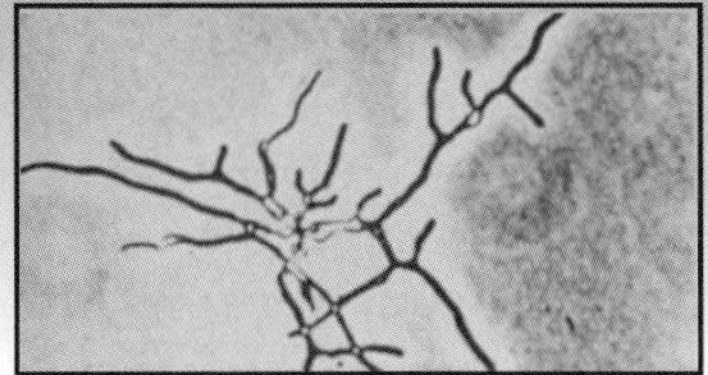

Fungal Basics

I Fungal Characteristics

Fungi **(molds, yeasts,** and **mushrooms)** are **eukaryotic organisms.** They **live on organic material** and are major recyclers. Spores of common environmental fungi are present in our air with much higher levels (a concern to compromised people) in moldy ceiling tiles, compost piles, and wet cellulosic carpet pads or sheet rock after floods or water damage.

Fungal cells have **two major chemical differences from human cells:**

A. ERGOSTEROL, not cholesterol, **is the major fungal membrane sterol.** Most antifungal agents exploit this difference.

1. **Imidazoles inhibit synthesis of ergosterol.** These are all **fungistatic.**
2. **Amphotericin B and nystatin (polyenes)** bind to cholesterol but **more avidly to ergosterol** disrupting fungal membrane integrity causing cellular contents to leak out. They **are fungicidal.**

B. FUNGI have complex carbohydrate and glycoprotein cell walls (notably chitin, glucans, and mannans with mannoproteins and fibrillar proteins), which are potential drug targets and which **stain with calcofluor-white fluorescent stain.** (Human cells do not fluoresce.) **Echinocandins are the fungal equivalent of β-lactams, inhibiting fungal glucan synthesis.** Capsofungin is the only licensed echinocandin; it is effective against *Candida* and *Aspergillus*.

II Fungal Structures

Fungi consist of **filaments (hyphae)** or **single-celled budding organisms (yeasts); some (dimorphic) species can convert from hyphal forms to yeast forms.**

A. HYPHAE (filamentous cells) grow to **form a mat called a mycelium.** Hyphae grow apically so dermatophytic cutaneous lesions should be sampled at the leading edge. Hyphae also make up fruiting bodies that we call mushrooms.

1. **Nonseptate hyphae lack regular cross walls** (septae), are irregularly **broad** (3–15 mm wide), and branch at obtuse angles. Nonseptate fungi like *Mucor* and *Rhizopus* cause rhinocerebral infections in IC or ketoacidotic diabetic patients; these infections progress rapidly to death.
2. **Septate hyphae have regular cross walls** and are **more uniform in width** (2–5 mm in diameter). **Most hyphae are septate and colorless (hyaline);** some genera's hyphae are septate and **dark (dematiaceous),** usually brown to gray.

B. YEASTS are oval to spherical cells that replicate by budding.

C. THE THERMALLY DIMORPHIC FUNGI are generally found **at room temperature or in the environment as filamentous fungi** (Ⓜ cold/mold) but grow **in the body as yeast or yeast-like form.** Important dimorphic fungi are *Histoplasma, Blastomyces, Coccidioides,* and *Sporothrix.*

D. PSEUDOHYPHAE are formed by ***Candida albicans*** when buds remain attached and elongate; they look **like hyphae with constrictions at each cell-cell juncture.** When incubated at 37°C (98°F) in rich medium, *Candida albicans'* yeast cells produce sprout-like projections called germ tubes as part of the conversion to hyphal forms. *Candida albicans* colonizes surfaces primarily with yeast forms; when it invades, yeasts, pseudohyphae, and true hyphae are seen.

III Fungal Sporulation

Most medically important fungi sporulate. Spore types (all asexual) include the following:

A. BLASTOCONIDIA are the **buds on yeasts.**

B. CONIDIA are asexual spores borne on the outside of sporulating hyphae. **Some** fungi **have both macroconidia** (slightly larger or multicellular) **and microconidia.** These are the primary **airborne spores** not usually seen in tissue except when fungi grow on the surface of sinuses of immunocompromised patients.

C. ENDOSPORES are spores **produced inside large spherical structures called spherules** in tissues such as lung by ***Coccidioides immitis.***

D. ARTHROCONIDIA (or arthrospores) are formed by **fragmentation of hyphae.** Hyphae breaking up into arthroconidia are seen in desert sand contaminated with *C. immitis* or in skin infected with dermatophytes.

IV Fungal Diseases

A. FUNGAL INFECTION. Fungal infections **(mycoses) may start from overgrowth of normal flora** (generally yeasts or dermatophytes), **from inhalation of fungal spores** (often from a dusty environment), **or from traumatic implantation** of spores into tissues. Fungal infections are particularly severe in compromised patients, often disseminating into the bloodstream (fungemia).

B. FUNGAL TOXINS

1. **Mycotoxicosis** is a poisoning from ingestion of fungal toxins produced in food, most notably the carcinogen aflatoxin in peanuts.
2. **Mycetismus** is illness resulting from ingestion of toxic mushrooms.

C. ALLERGIC REACTIONS

1. **Allergy** to fungi plays a role in **allergic bronchopulmonary aspergillosis** and **farmer's lung.**
2. **Sick building syndrome** arises from **inhalation of volatile fungal toxins and spores,** which aggravate allergies.

V Laboratory Identification

A. FUNGAL CULTURE. Special fungal media are used.

1. **Sabouraud's agar,** a standard fungal medium on which most fungi will grow, is used to grow yeasts and **as a clue to fungal infection on exams.**
2. **Three media**—fungal versions of **blood agar, blood agar with antibiotics** (to inhibit any potential bacterial contaminants), **and blood agar with antibiotics and cycloheximide** (to inhibit fungal contaminants but also some opportunistic or pathogenic fungi)—are generally used in culturing suspected systemic fungal infections.
3. **Identification of fungal isolates** is done by morphology, biochemical tests (yeasts), immunologic tests, or genetic probe.

B. MICROSCOPY. Fungi are visualized in tissues using the following methods:

1. **Skin scrapings are dissolved in 10% KOH** to improve ability to see the fungi with a light microscope.
2. **Calcofluor white binds to the complex carbohydrate cell wall and lights up the fungi a bright blue white;** human cells do not fluoresce.
3. **Special fungal stains** include the **silver stain** (**fungi** and basement membranes stain **silver**), and periodic acid Schiff **(PAS)** reaction **(fungi stain hot pink).**
4. **India ink** (used as wet mount medium for CSF sediment) **highlights the capsule of *Cryptococcus neoformans*,** but has a sensitivity of only 50% so it cannot rule out cryptococcal meningitis. Immunologic methods, such as latex particle agglutination (LPA) to identify the presence of the capsular polysaccharide antigen in CSF, are much more sensitive than India ink.
5. **Immunofluorescent stains** are available for identification of some fungi in tissues.

C. ANTIBODY OR ANTIGEN DETECTION. Tests for detection of patient antibodies or fungal antigens in body fluids are available. The most critical of the latter group is the detection of **Cryptococcal polysaccharide in CSF,** as mentioned above.

D. SKIN TESTING. This only demonstrates exposure.

Chapter 27

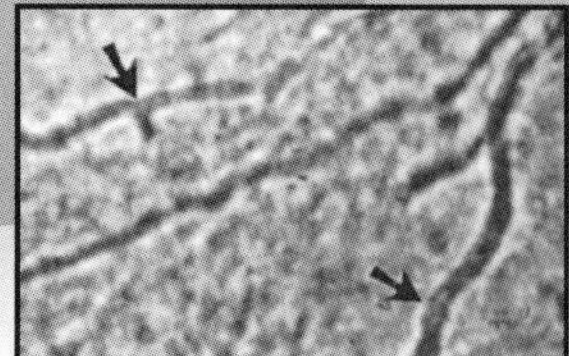

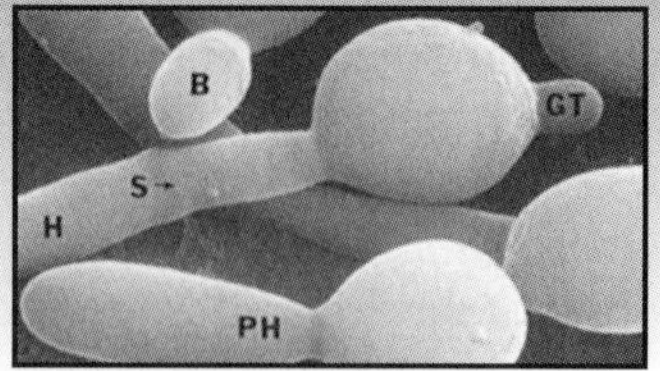

Fungi That Cause Skin and Subcutaneous Infections

Malassezia, Dermatophytes *(Trichophyton, Microsporum, Epidermophyton), Candida, Sporothrix*

I *Malassezia*

Malessezia furfur, a **lipophilic "yeast"** found on skin as **normal skin flora,** can cause the following disorders:

A. Pityriasis or tinea versicolor, which presents as **hypopigmented spots** on the chest, back, or both. Patients usually present with complaints of speckled tanning. Skin scrapings contain **clusters of oval cells** and **short, curved, septate hyphae.** Ⓜ TV (a colorful disease) with your TV dinner of spaghetti and meatballs and, at your feet, your "fur ball" (*Malessezia furfur* [my next dog's name!]).

B. Fungemia in premature infants on intravenous lipid supplements. *M. furfur* grows on blood agar only if a thin coat of sterile olive oil is added.

II Dermatophytes

A. THREE SPECIFIC GENERA of these filamentous fungi infect skin and other keratinized tissues:

1. ***Trichophyton*** **infects skin, hair, and nails.**
2. ***Microsporum*** **infects hair and skin.**
3. ***Epidermophyton*** **infects nails and skin.**

B. EPIDEMIOLOGY. Dermatophytes are **zoophilic, anthropophilic** (from other humans), or **geophilic. Highly inflammatory dermatophyte** infections (tineas) **are usually acquired from animals** (e.g., farmers often contract it when setting up milking machines). **Anthropophilic** dermatophytes are spread by **fomites** (e.g., hats, combs, and shower room floors) and by **direct person-to-person contact.** Many dermatophytes are quite contagious.

C. INFECTIONS (BY LOCATION). Most tineas or ringworms are named by location: **tinea capitis** (hair and scalp), **tinea barbae** (bearded region), **tinea corporis** (glabrous skin), **tinea cruris** or **"jock itch"** (groin and perineal area), and **tinea pedis** ("athlete's foot"). The etiologic agents for either tinea cruris or tinea pedis are most commonly *Epidermophyton floccosum* and *Trichophyton* spp. **Favus or tinea favosa** (named for the honeycomb scutula or crusts) is most serious, **causing permanent scarring and scalp hair loss.**

D. DIAGNOSIS is either **clinical or by microscopy** (KOH mounts of skin, hair, or nail show arthroconidia and hyphae).

E. DERMATOPHYTID ("id") reaction is an allergic response to circulating dermatophyte antigens released by dying fungi during antifungal treatment.

III *Candida*

Candida is a genus of yeasts. *Candida albicans,* which forms both **pseudohyphae and true hyphae in tissue,** is common.

A. EPIDEMIOLOGY. Many species of *Candida* are found as normal flora of the mucous membranes and skin. Predisposing conditions, such as continuous moisture, occluded skin surfaces, antibiotic use, or diabetes, may cause mucocutaneous overgrowth, resulting in symptoms.

B. INFECTION: CANDIDIASIS. *Candida* spp. infections range from minor but painful mucocutaneous lesions (nail infections, diaper rash, yeast vaginitis) to major problems, such as septicemia or cerebritis in immunocompromised patients and catheter infections.

1. **Diaper rash and yeast vaginitis** are characterized by pain and erythema, with a sharp margin between affected and normal skin. Satellite lesions may be present. Non-*albicans* vaginitis is more likely to recur and to have azole resistance. Diagnosis: saline wet mount; amine test (negative)
2. **Thrush, characterized by a "cheesy" white coating of the oral mucosa with a painful erythematous base,** occurs in premature babies and immunocompromised patients.

IV *Sporothrix*

Sporothrix schenckii is a dimorphic fungus.

A. EPIDEMIOLOGY. *S. schenckii* grows as a filamentous fungus on a variety of plant materials. When traumatically implanted into human tissues, the organism converts into a cigar-shaped yeast. The implantation may involve thorns, floral wire, or slivers; commonly involved plants include roses, plum trees, or sphagnum moss.

B. INFECTION. **Sporotrichosis** is commonly called **rose gardener's disease.** There may be **solitary subcutaneous lesions at the trauma site** (usually on extremities), or there may be lesions along the lymphatics **(lymphocutaneous sporotrichosis),** with lymph nodes farther from the trauma progressively less involved.

C. DIAGNOSIS. Yeast cells (cigar-shaped to quite spherical) are difficult to find in tissue. Cultures (filamentous cells in 30°C culture) should be done.

D. TREATMENT. Oral potassium iodide (no antifungal activity) helps break down the granulation tissue, so that the immune system cells can get in to kill the *Sporothrix*. Itraconazole is also used.

Chapter 28

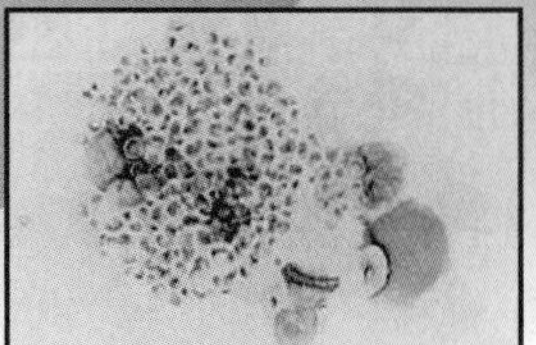

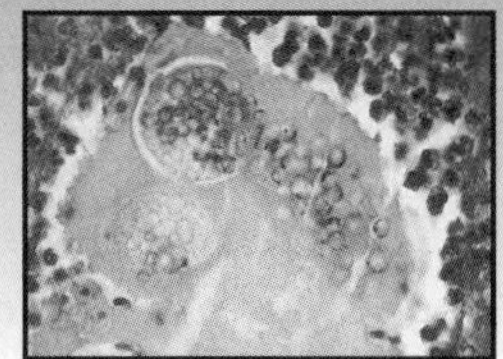

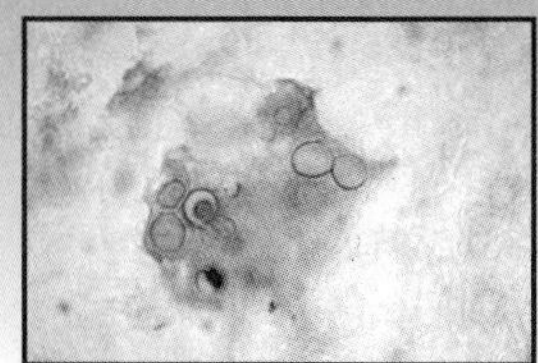

Systemic Fungal Pathogens

Histoplasma, Coccidioides, Blastomyces

I Introduction

Histoplasma, Coccidioides, and ***Blastomyces*** are the **major systemic fungal pathogens in the United States.** Some generalizations can be made about all three genera:

A. **ALL ARE THERMALLY DIMORPHIC.**

B. **INFECTION** starts with **inhalation of spores from an environmental source,** often in "dust." **Infection is not transmitted human to human.**

C. ***HISTOPLASMA, COCCIDIOIDES,*** **AND** ***BLASTOMYCES*** cause **asymptomatic or acute self-resolving fungal pneumonias in 95% of cases.** The remaining 5% become chronic or disseminate to other tissues.

D. **THE ACUTE INFECTIONS,** which commonly heal without treatment in a healthy non-compromised person, may reactivate later under immunocompromising conditions (similar to tuberculosis).

II *Histoplasma capsulatum*

A. ***HISTOPLASMA CAPSULATUM*** is **thermally dimorphic** and has **no capsule.**
   1. **The environmental form of *H. capsulatum* is filamentous** fungus **with** small **microconidia** and **large tuberculate macroconidia** (spherical, with short finger-like projections).
   2. **The tissue form** of ***H. capsulatum*** is a **facultative intracellular yeast** seen as small oval budding yeasts (2–4 μ) **inside cells of the reticuloendothelial system** (RES).

B. EPIDEMIOLOGY. ***H. capsulatum*** is associated with **bird- and bat-enriched soil.** Endemic areas are the **great plains around the Ohio, Mississippi, and Missouri rivers.** Typical exposure comes from **spore inhalation during dusty activities,** such as cave exploring, demolition work, or cleaning old chicken coops.

C. HISTOPLASMOSIS

1. **Primary histoplasmosis ranges from asymptomatic to acute self-resolving fungal pneumonia,** known as the fungus flu. Symptoms are usually cough, fever, malaise, weakness, chest pain, headache, myalgia, chills, nausea, anorexia, and weight loss. Radiographs in sicker patients show pulmonary infiltrates, with hilar lymphadenopathy being common. Lesions have a tendency to calcify, especially in the young.
2. **Systemic infections occur in immunocompromised patients,** including AIDS patients. Some infections may be recrudescences of earlier "healed" infections. **Mucocutaneous lesions (oral or genital) are common** in patients with disseminated disease.

D. LAB ID. Diagnostic suspicion should arise if the patient **has pneumonia but has not responded to antibacterial drugs and has had exposure to a dusty environment in the endemic area 1 to 2 weeks earlier.** Sputum is rarely positive for yeasts, but **Giemsa-stained smears and cultures of peripheral blood, bone marrow, or urine may be positive for *H. capsulatum*.** A fourfold increase in serum antibody titers from acute to convalescent stage is diagnostic.

Coccidioides immitis

A. COCCIDIODIES IMMITIS is a **thermally dimorphic pathogen.**

1. **Environmental form.** ***C. immitis*** **grows in sand** as **hyphal filaments** that develop into **arthroconidia.**
2. **Tissue form. Inhaled arthroconidia develop into spherules** (30–60 μ), which produce **endospores** (2–5 μ) internally.

B. RESERVOIR AND TRANSMISSION. ***C. immitis*** **is found in desert sand in the southwestern United States.** Airborne arthroconidia are inhaled when sand is disturbed.

C. COCCIDIOIDOMYCOSIS

1. **Primary coccidioidomycosis (Valley fever)** ranges from **asymptomatic infection to self-limited fungal pneumonia.** It is most common in children and newcomers to the endemic area. Symptomatic cases present with cough, fever, and dull chest pain, along with flulike symptoms resembling any atypical pneumonia. Erythema nodosum is a good prognostic sign.
2. **Disseminated form.** Coccidioidomycosis disseminates more commonly in racial groups with certain HLA types (particularly African Americans and Filipinos), in women in the third trimester of pregnancy, and in immunocompromised patients. It most commonly **disseminates to skin, subcutaneous tissue, bones, joints, or meninges.**

D. LAB ID. Incubation is about 4 weeks. Cultures are hazardous. **Microscopy** on sputum, urine, or bronchial washings **may show spherules.** A fourfold increase in serum antibody titers from acute to convalescent stage is diagnostic.

IV *Blastomyces dermatitidis*

A. *BLASTOMYCES DERMATITIDIS* is a **thermally dimorphic fungus.**

1. **The environmental form** is **hyphae** with **conidia arising** off short, lateral "stalks."
2. **The tissue form** is a large (8–15 μ) **yeast** with a **broad-based bud** (i.e., the juncture between the cells is wide) and a **thick, double, refractile cell wall.**

B. EPIDEMIOLOGY. *Blastomyces* is found in nearly the same geographic region as *Histoplasma*, as well as in the mid-Atlantic region of the United States and northern Minnesota. Environmental association is uncertain but is probably rotting wood. Conidia are inhaled.

C. BLASTOMYCOSIS. This fungal pneumonia is less common than histoplasmosis but more likely to have an indolent onset. It is also less likely to self-resolve and more likely, when it does disseminate, to involve skin and bone.

D. LAB ID. Cultures are usually done by reference labs. Biopsy specimens examined by microscopy may show large yeast with a wide, budding base and a double, refractile wall. (M Blasto: broad-based budding yeasts with big [thick] wall.)

Chapter 29

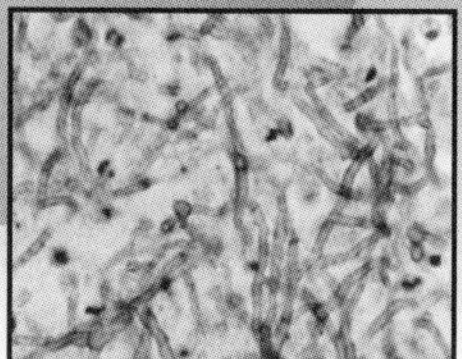

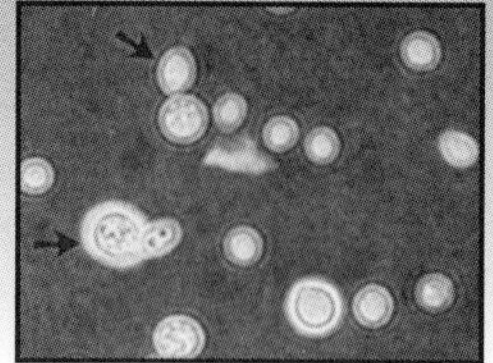

Opportunistic Fungi

Aspergillus, Candida, Cryptococcus, Mucor, Rhizopus, Absidia, Pneumocystis

I Introduction

The major **opportunistic fungi are monomorphic**, except for the polymorphic *Candida*. The opportunists are either fairly common environmental fungi of low virulence or normal epithelial or mucous membrane flora (*Candida*).

II *Aspergillus fumigatus*

Aspergillus fumigatus is a **monomorphic filamentous** fungus with dichotomously **branching hyphae that have acute (<45°) angles** and **small, airborne conidia.**

A. EPIDEMIOLOGY. Fungi of the genus ***Aspergillus* are ubiquitous** as **major recyclers;** as such, they are found **in and on almost any moldy organic material** (food, **ceiling tiles** that have been wet, **compost,** and so on). Inhaled airborne **conidia of *A. fumigatus* are small enough to reach the alveoli.** In normal hosts, spores are removed by the mucociliary elevator, or their growth is controlled by macrophages and neutrophils.

B. DISEASES

1. **Allergic bronchopulmonary aspergillosis** occurs in people with asthma or allergies. Because the *Aspergillus* may grow on bronchial mucus plugs, the allergen is ever-present. Immunoglobulin E (IgE) levels will be high.
2. **Fungus balls grow in pre-existing lung cavities,** but **hyphae do not penetrate the tissue.** Disease is characterized by cough, sometimes accompanied by hemoptysis and high IgE, as antigen is in the lung.
3. **Farmer's lung** is a complex allergic disease that includes allergies to *Aspergillus* and other thermophilic bacteria and fungi found in silage.

4. **Invasive aspergillosis** (most commonly pneumonia) occurs under conditions of **severe neutropenia, chronic granulomatous disease, cystic fibrosis,** and **burns. Sinus surface colonization** with sporulation or direct **inhalation of large numbers of spores from the environment** leads to pneumonia. ***A. fumigatus*** **may spread to the brain by direct extension** from sinuses by **hematogenous spread.** In burn patients, *Aspergillus* **cellulitis** may also disseminate.

III *Candida*

Candida **spp.** are **polymorhphic fungi** with **yeasts and pseudohyphae;** a few, including ***C. albicans,*** also produce **germ tubes and true hyphae** in tissues.

A. **EPIDEMIOLOGY. *Candida*** **spp.** are found as **normal mucocutaneous flora** but, under certain conditions, may overgrow and invade.

B. **DISEASES: CANDIDIASES**
   1. **Oral thrush** occurs in premature infants, patients on antibiotics, and immunocompromised (IC) hosts. **Oral thrush** may progress **to esophagitis, then gastritis,** and ultimately, **through bowel defects, septicemia.** Denture stomatitis often involves *Candida.* **Perlèche** (soreness in the mouth angles) suggests **malnutrition.**
   2. **Mucocutaneous candidiasis. Yeast vaginitis** is a recurrent problem for some women. It is more common in diabetics (high sugar appears to promote the growth and adherence) and in HIV-positive women as CD4+ get lower. **Discharge is curdy, white,** and **adherent** with **normal pH** and negative whiff test (no amine smell when mixed with KOH). **Diaper rash:** damaged diaper-area skin, in infants and toddlers encourages colonization with *Candida* spp. **Moist, occluded skin folds in anyone** are susceptible to candidiasis, with risk increased if there are endocrine disturbances or damage to the skin (radiation, diabetes, obesity, etc.).
   3. **Candidemia (septicemia or fungemia)** is usually found in **patients with catheters, with immunocompromising conditions, with burns,** or **who are postsurgical.**
   4. **Endocarditis (with transient fungemias)** occurs in **intravenous drug abusers** or people with **indwelling catheters.** Stroke syndrome, from vegetation pieces breaking off, may occur.
   5. **Chronic mucocutaneous candidiasis (CMCC).** Patients (often with endocrine problems) who are **anergic** (T cell defects?) develop CMCC. Successful imidazole treatment may allow immunity to develop.

C. **C. *ALBICAN'S* VIRULENCE** is based on the ability of **yeast cells to bind human cells** and **extracellular matrix** and **to produce proteases and phospholipases to invade tissue.** Normal skin and a healthy immune system usually control infection. However, damage to skin or mucosa, particularly along with a compromised immune system, leads to invasion.

D. **LAB ID. Microscopy (pseudohyphae, true hyphae, and budding yeast cells)** is used in diagnosing superficial infections. Systemic infections are cultured and speciated using germ tube formation and chemical tests.

E. **TREATMENT.** Antifungal susceptibility is increasingly important because **non-*C. albicans* strains have increasing drug resistance.** This is a major problem, as these strains are often in IC patients. Most *Candida* respond to **polyenes, capsofungin, and imidazoles.** (Important exceptions: *C. krusei* is resistant to fluconazole; *C. parapsilosis* is resistant to voriconazole.)

 *Cryptococcus neoformans*

Crytococcus neoformans is a **monomorphic yeast** and the **only fungal opportunist/pathogen with a** polysaccharide (or any) **capsule** in vivo.

A. EPIDEMIOLOGY. The environmental source is **soil enriched with pigeon droppings.** The organism is inhaled. Exposure may be common.

B. CRYPTOCOCCOSIS. Except in IC patients, infection is restricted to the lungs.

1. **Pulmonary infection** is asymptomatic in the majority of cases, but some immunocompetent patients may have severe pneumonia with mucoid sputum.
2. **Cryptococcal meningitis** is the **dominant meningitis in AIDS patients;** it is also seen in patients with cancer (e.g., Hodgkin lymphoma). Pulmonary signs are often absent.

C. LAB ID. Diagnosis of meningitis is made using cerebrospinal fluid **(CSF). Protein will be elevated** and **glucose will be <40 mg per dL.**

1. **The latex particle agglutination** (LPA) test, which looks for **capsular antigen (polysaccharide)** in the CSF, is rapid and sensitive.
2. **Microscopy on India ink wet mount** of CSF sediment, looking for **budding yeasts with capsular "halos,"** is rapid and useful *if* the result is positive. This test **misses 50% of culture- or LPA-documented cases** of cryptococcal meningitis and so is a rule in, not out, test.
3. **CSF cultures** are done but *C. neoformans* is slow to grow.

 *Mucor, Rhizopus, Absidia* (Zygomycota)

The genera of ***Rhizopus, Mucor,*** and ***Absidia*** consist of **nonseptate filamentous fungi** (Zygomycota).

A. EPIDEMIOLOGY. These **nonseptate, monomorphic, filamentous fungi,** which are found in the **environment** (e.g., **soil, strawberries, and moist bread**), produce airborne spores that appear to colonize sinus tracts.

B. RHINOCEREBRAL INFECTION (mucormycosis, phycomycosis, or zygomycosis) presents in **ketoacidotic diabetic** patients and **cancer** patients. Symptoms include **paranasal swelling, mental lethargy, and hemorrhagic exudates in the nose and sometimes the eyes.** These nonseptate fungi (sometimes called *bread molds*) are rapidly invasive (without respect for barriers) from the sinuses into the brain, penetrating vasculature and causing hemorrhage.

C. LAB ID involves immediate, or stat, examination of a KOH stain of necrotic tissues, looking for **broad, ribbonlike, nonseptate hyphae with about 90° angles on branching.**

D. TREATMENT includes **rapid debridement of necrotic tissue, amphotericin B,** and, **if diabetes is involved, lowering blood glucose.** The **fatality rate is high.**

Pneumocystis jiroveci

Pneumocystis jiroveci (formerly *P. carinii*) is an **obligate extracellular fungus** that **cannot be grown** in the lab. As a result of ribotyping* and other molecular techniques, *Pneumocystis* is now considered a fungus.

A. DISEASE. *Pneumocystis* **exposure is quite common,** but **disease occurs primarily in AIDS patients** or in severely **malnourished or premature infants. Pneumocytis pneumonia (PCP)** presents as an **interstitial pneumonia,** which, in radiographs, may have a ground-glass appearance.

B. LAB ID. Pneumocystis cannot be cultured. Lung tissue stained with H & E shows alveoli with a foamy appearance, sometimes referred to as a honeycomb appearance. With **silver stain, silver-stained cysts** appear in the center spaces. **Fluorescent antibody** and **polymerase chain reaction** (PCR) are available for diagnosis of PCP.

C. TREATMENT. Trimethoprim/sulfamethoxazole prophylaxis is used in known HIV+ patients with CD4+ counts <200 cells per mm^3.

*Ribotyping compares ribosomal base sequences, which are very well conserved within groups of related organisms.

Part **V**

Parasites

Chapter 30

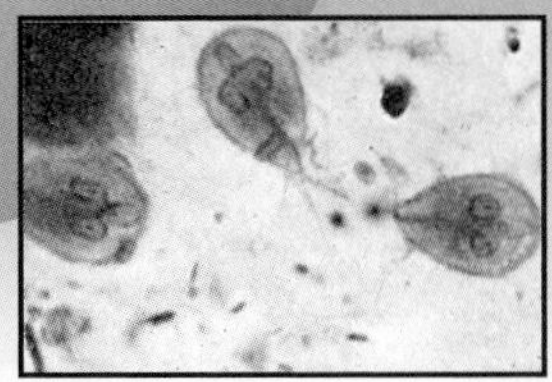

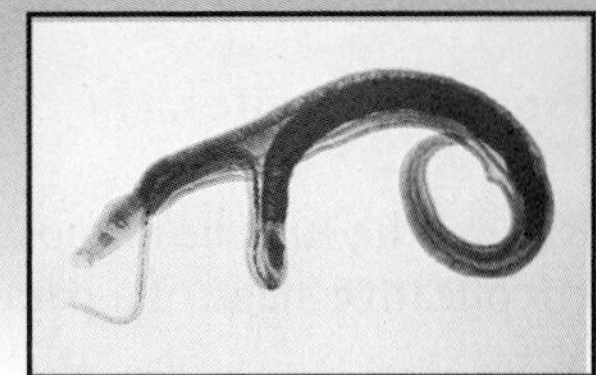

Parasite Basics

I Overview of Parasites

A parasite is an organism (from virus to animal) that **lives in or on another organism (the host) and does some damage to the host** in the process. Chapters 30 through 34 discuss animal (protozoan or helminthic) parasites. **Animal parasites are eukaryotic organisms with no cell walls, ranging from single-celled organisms to large, multicellular worms.** There are several types of parasites:

A. A FACULTATIVE PARASITE (or free living) can live in association with its host or separately.

B. AN OBLIGATE PARASITE cannot live free of the host for at least some stage of the life cycle.

C. A PARASITE with a **complex life cycle requires more than one host.**

II Hosts

Hosts are the organisms that provide nutrition and a **place for the parasite to replicate.**

A. A RESERVOIR HOST maintains a parasite and may be the source for human infection. **An essential reservoir host is one without which the parasite cannot exist** (e.g., cats for *Toxoplasma*).

B. AN INTERMEDIATE HOST either maintains the asexual stage(s) of a parasite or allows development of the parasite to proceed only to the larval stage.

C. A DEFINITIVE HOST is **one in which the adult or sexual parasites develop.**

Vectors

Vectors are biological systems that spread parasites.

A. A BIOLOGICAL VECTOR (e.g., the *Anopheles* mosquito in malaria) **serves both as a vector and a host** for the replicative stage of a parasite.

B. A MECHANICAL VECTOR transmits a parasite without being a host (e.g., flies "tracking" *Chlamydia trachomatis* from one eye to the next).

IV Major Groups of Animal Parasites*

A. PROTOZOA are single-celled animals. Trophozoites are the more delicate (i.e., don't survive stomach acid) **motile forms of protozoa. Cysts,** with their chitinous coatings, **are the more inert but infectious protozoan stage in fecal-oral transmission.** Cysts **survive in the environment; stomach acid aids in their excystment.**

1. **Amebas move by pseudopodia.** The granular endoplasm (involved in the nutrition and synthetic processes) is strikingly differentiated from the clear ectoplasm (involved in rhizopod formation). Major pathogenic amebas include *Entamoeba histolytica, Naegleria fowleri,* and *Acanthamoeba* spp. *E. dispar,* which cannot be reliably distinguished visually from *E. histolytica,* is a nonpathogen even in AIDS patients; the antigen testing distinguishes the two.
2. **Flagellates move by one or more flagella.** Pathogens include *Giardia lamblia, Trichomonas vaginalis, Trypanosoma* spp., and *Leishmania* spp.
3. **Sporozoans (Apicomplexa) are obligate intracellular protozoans with complex life cycles and at least two different hosts.** Sporozoans include *Plasmodium, Cryptosporidium parvum,* and *Toxoplasma gondii.*

B. ROUNDWORMS (NEMATODES) are worms that are round in cross section, with tapering ends. (Ⓜ A nemesis is someone who bugs you and is always "a-round" you. Roundworms are nematodes.) Nematodes have **separate male and female individuals** and well-developed gastrointestinal (GI) tracts. "High-yield" nematodes are *Ascaris,* hookworms, and *Enterobius.* Roundworms infect several billion of the world's population.

C. FLATWORMS are flat, multicellular worms. There are two groups:

1. **Trematodes,** also called flukes (Ⓜ Think: fluke, flutter, tremor, trematode), **are generally flat, fleshy, nonsegmented worms.** All **trematodes have complex life cycles involving snails as intermediate hosts** and **water transmission** to humans. **Schistosomes are important flukes.**

*Major groups are considered important for physicians, because antiparasitic agents are often broad enough to be used against a whole group (e.g., an antinematode drug).

2. **Cestodes are tapeworms (segmented flatworms).** Cestodes have complex life cycles with **at least two hosts. Intermediate hosts ingest eggs that develop into larvae in tissue, often resulting in serious disease. Definitive hosts ingest larvae** from infected tissues. The larvae **develop into adult tapeworms in the intestines of the definitive host.**

V Characteristics of Parasitic Disease

A. **SYMPTOMS are usually proportional to parasite burden** but **may become more severe** once the **individual is sensitized** to parasitic components.

B. **REINFECTIONS may occur from environment or by autoinfection** (autoinfection of the host without the organism going through developmental stages elsewhere is common with *Strongyloides* and *Enterobius*).

C. **CHRONIC INFECTIONS** (e.g., Chagas disease) **may occur,** with or without overt acute disease. **Some parasites can remain viable in humans for long periods, in some cases throughout a person's life, and may cause periodic disease symptoms.**

D. **IMMUNOCOMPROMISING CONDITIONS may cause reactivation of latent infections** (such as toxoplasmosis), **increased susceptibility,** and, **often, more severe disease.**

E. **EOSINOPHILIA** is sometimes a **clue to parasitic disease,** although it usually only occurs during **larval worm migration through the vasculature or tissue.**

VI Lab Diagnostic Methods

Lab diagnostic methods for parasitic infections vary with frequency of infections in a country and the size and expertise of the laboratory. The more sophisticated tests are often available only in larger reference laboratories.

A. **MICROSCOPIC EXAMINATION of tissues, blood, and feces** (usually following concentration methods) is still used. In the United Staes, **worms infecting the GI tract are still mostly identified** by microscopic identification of the **eggs in feces.** Helminthic eggs likely to be used as clues in USMLE cases are shown in Figure 30-1.

B. **CULTURES.** Culture techniques are used to identify some protozoans.

C. **ANTIGENS/ANTIBODIES.** Tests demonstrating the presence of antigens for a specific organism are now available for some tissue, blood, and GI tract pathogens. Tests demonstrating patient antibody are also available for some parasites.

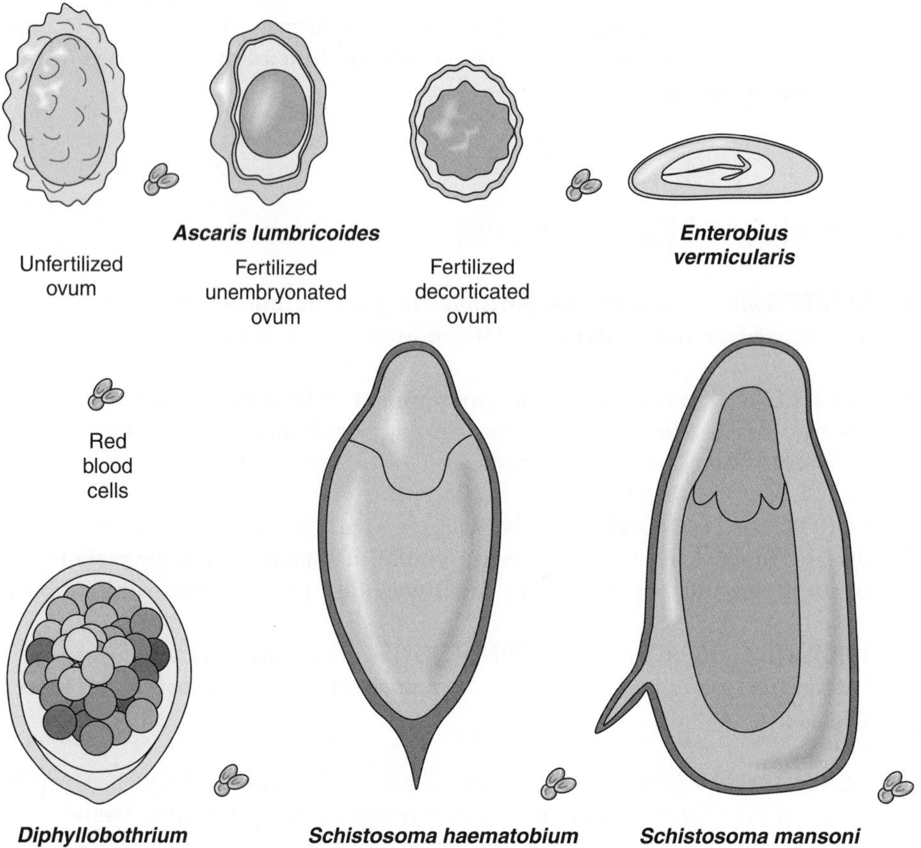

● **Figure 30-1** Common worm eggs and their proportional sizes.

Chapter 31

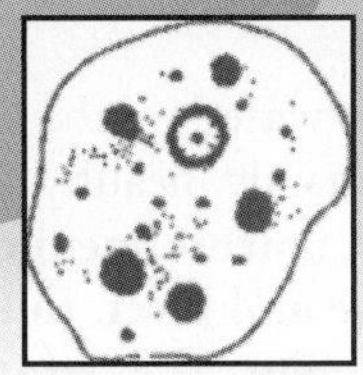
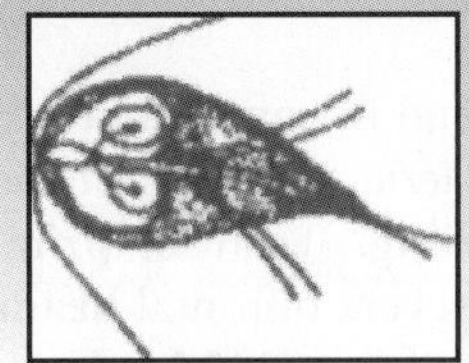
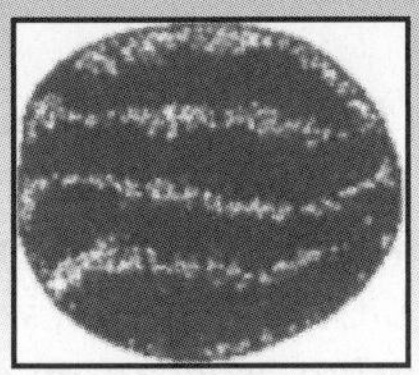

Protozoan Parasites

Entamoeba, Naegleria, Acanthamoeba, Giardia, Trichomonas, Trypanosoma, Leishmania, Cryptosporidium, Plasmodium, Toxoplasma

I Amebas

A. ***ENTAMOEBA HISTOLYTICA*** is a human pathogen (second in mortality to malaria and schistosomiasis) of the large intestine that is **spread by cysts** via the fecal-oral route in areas of poor sanitation or by oral anal sex. Whether the person gets ill depends on underlying health and nutrition.

1. **Amebic dysentery. The infective cysts** excyst (with the aid of stomach acid) to release the ameboid trophozoites, which adhere to colonic mucosa via galactose-specific lectins. The trophozoites replicate and invade the large intestinal mucosa in crypts, causing **inverted flask-shaped ulcers,** and feed on red blood cells (RBCs). **Dysentery symptoms** include **high fever** with **intermittent crampy diarrhea** or **dysentery** (abdominal cramps with tenesmus and blood and pus in stools), **alternating with normal stools or constipation. Extraintestinal abscesses** (particularly in **liver and lung** and sometimes referred to as *extraluminal giardiasis*) are a **common complication** of *E. histolytica.* As the trophozoites leave the gut, they encyst. An infected person may excrete 15 million cysts per day. What were once thought to be inapparent infections with nonvirulent *E. histolytica* are now thought to be colonization with ***E. dispar*, a nonpathogen** with identical appearing cysts. ***Entamoeba coli*, another nonpathogen** colonizing the intestines, does not have the "wagon wheel" nuclei (see below).
2. **Lab ID.** Microscopically, the ***E. histolytica* trophozoite** can be recognized by its **"wagon wheel" nuclei (sharp central dot with fine chromatin "spokes")** and **its ingested RBCs,** which distinguish it from the commensal *Entamoeba coli,* which does not cause disease. A new **stool antigen test** improves diagnosis and distinguishes nonpathogenic *E. dispar* from pathogenic *E. histolytica.*

In extraintestinal disease (i.e., liver ulcerations), serology is used to make a diagnosis.

B. ***NAEGLERIA*** **is a free-living ameba** found in **contaminated hot water** (in the United States, mainly in the southeast). *Naegleria* infects humans (mainly healthy young males) during **swimming or water skiing.** It causes **primary amebic meningoencephalitis (PAM).** Onset is rapid, with severe bifrontal **headache and fever, followed by nausea, vomiting, sensation of constant odd odor or taste, and irrational behavior.** There are meningeal, frontal, temporal, and cerebellar signs. Coma and death are common within 6 days. On microscopy, the cerebrospinal fluid **(CSF) should show trophozoites, polymorphonuclear neutrophil leukocytes** (PMNs), **RBCs, elevated protein**, and **low glucose.**

C. ***ACANTHAMOEBA*** **is a free-living soil ameba (resistant to desiccation)** that is probably spread by inhalation of cysts. *Acanthamoeba* diseases include:

1. **Granulomatous amebic meningoencephalitis (GAE) in immunocompromised (IC) patients.** GAE has a more chronic onset (less fulminant and more benign) than PAM but still progresses to death. Symptoms reflect focal lesions. Spiky (wrinkled) trophozoites are seen in CSF or brain tissue.
2. **Keratitis.** ***Acanthamoeba*-contaminated homemade contact lens saline,** along with minor eye abrasion, can cause **keratitis,** which is characterized by **corneal inflammation and pain.** It can cause **blindness.**

II Flagellates

A. ***GIARDIA LAMBLIA*** is a distinctive **teardrop-shaped flagellated protozoan** with a **ventral sucking disk. Cysts** are the **infectious form.**

1. **Epidemiology.** *Giardia lamblia* is found **worldwide** in water contaminated by infected beavers, muskrats, or areas of poor human sanitation. (Expect cases in **wilderness campers.**) It is transmitted by **ingestion of cysts** in untreated lake or river water, fecally contaminated food (poor general sanitation or day-care diaper changing), or oral-anal sex.
2. **Giardiasis.** *Giardia*'s **ventral sucking disk** attaches to the **duodenal/jejunal surface, blocks adsorption,** and causes irritation and **blunting of microvilli. Symptoms vary** from mild **watery diarrhea** to **malabsorption syndrome with epigastric pain, severe abdominal cramping, flatulence, and light-colored, fatty stools.** Restricted surface for release of lactases in severe cases leads to **temporary lactose intolerance.**
3. **Lab ID.** Because of its attachment, *Giardia* is not seen in the feces of 60% of those infected. ***Giardia*** **fecal antigen tests** have replaced the **string test** when microscopy is negative.

B. ***TRICHOMONAS VAGINALIS*** is a **contact** (commonly **sexual) transmitted flagellate** with both an **undulating membrane** and **a polar tuft of flagella.** The **infectious trophozoite** has a **contact-dependent cytopathic effect.**

1. **Trichomonal vaginitis** is characterized by a **malodorous (fishy), purulent yellow (60%) or frothy (10%) vaginal discharge** and **erythematous vaginal mucosa. Dyspareunia, urinary frequency,** and **dysuria** may also be present. Males are often asymptomatic.

2. **Lab ID. Vaginal discharge pH is ≥4.5; discharge in KOH often gives amine odor;** in saline, it should show numerous **PMNs** and **motile/jerky trophozoites.** Microscopy has sensitivity of 50% to 60% (i.e., negative in 40% to 50% of infected people). New tests **(immunofluorescence, antigen, genetic probe) are more sensitive.**

C. *TRYPANOSOMA BRUCEI RHODESIENSE AND T. BRUCEI GAMBIENSE* (*T. brucei*) are **flagellates** spread by the **tsetse fly** in **Africa.** There are 45,000 reported cases per year, but with estimates of up to 5×10^6 infected.

1. **Trypanosomiasis, or African sleeping sickness (ASS).** Within 2 to 3 days, a red, tender nodule/chancre appears at the tsetse fly bite site. The *T. brucei* invade the blood and lymphatics, causing fever, headache, malaise, rash, and posterior cervical lymph node enlargement (Winterbottom sign). Symptoms recur with each wave of new antigenically different *T. brucei*. Within 3 to 6 weeks, the less common Rhodesian trypanosomaisis progresses to myocarditis and central nervous system (CNS) symptoms, with death in untreated cases coming in 6 to 9 months. Gambiense forms may have recurring fever and parasitemias for years before CNS invasion and then a slow decline to lassitude, coma, and death.
2. **Virulence and Lab ID.** The progressive nature of untreated ASS is partially due to **extreme antigenic variation** of trypanosomal **surface glycoproteins** (more than 1,000 variants), which allows evasion of the immune system. **Immunoglobulin M (IgM) levels are usually >4 times normal.** Normal serum IgM levels rule out ASS. Diagnosis is by the microscopic finding of trypanosomes.

D. *TRYPANOSOMA CRUZI* is another **hemoflagellate** that is found from Mexico to South America, where **American trypanosomiasis (Chagas disease)** is a significant cause of death.

1. **Epidemiology. Mammals are intermediate hosts,** and **reduviid bugs ("cone-nosed" or "kissing" bugs) are definitive hosts. Transmission may be through the placenta** (causing a high rate of stillbirths), **organ transplantation, blood transfusion** (an increasing problem in the United States), **or reduviid bugs that defecate as they bite.** The infected person scratches, depositing the fecal *T. cruzi* into the wound, where it is replicated extracellularly, producing a **painful, indurated ulcer (chagoma),** and then **spreads through the lymphatics and bloodstream.** The motile *T. cruzi* binds to mammalian cells (especially muscle and glial cells) through a surface protein (penetrin) and then invades, by developing into the **nonmotile intracellular form (the amastigote),** and replicates. The infected cells get huge with new organisms and lyse, spewing cells back into the bloodstream.
2. **Chagas disease or American trypanosomiasis. Spread of *T. cruzi*** leads to **additional chagomas,** often **swelling around one eye (Romaña sign); variable high fever; malaise; lymphadenopathy; and hepatosplenomegaly.** Spontaneous cure may occur.
   a. In the **very young, progressive disease** often leads to meningoencephalitis and death.
   b. In **older children or adults,** disease is more **chronic,** causing an **enlarged, flaccid heart** that may **cause sudden death** or, less commonly, **megaesophagus or megacolon.** Unfortunately, people may not recognize that they are infected, and diagnosis of chronic infections is difficult in nonendemic areas. Transfusions with infected blood to neonates or IC patients may lead to severe disease.
3. **Lab ID. Microscopy** may reveal the **C-shaped** (Ⓜ C is for Chagas and the shape of the extracellular *T. cruzi*), **flagellated trypomastigotes** in blood or nonmotile

amastigotes in tissue; culture, xenodiagnosis, enzyme-linked immunosorbent assay (ELISA), and polymerase chain reaction (PCR) are also available. (**Xenodiagnosis: Lab reduviids feed on patient** and are later dissected and examined for the trypanosomes.)

E. ***ALL LEISHMANIA*** spp. multiply in and are **transmitted by sand flies. *Leishmania* infects mammalian macrophages.** Leishmanial diseases vary with species:
   1. ***L. donovani*** complex causes **visceral leishmaniasis** (kala-azar), invading skin (early), spleen, liver, and bone marrow. Hepatosplenomegaly, anemia, weight loss, and indefinite feeling of unwellness are common. There may be fever, including double (dromedary) or triple fever peaks daily.
   2. ***L. tropica*** complex causes **cutaneous leishmaniasis,** also called Oriental sore.
   3. ***L. braziliensis*** complex causes **mucocutaneous leishmaniasis** and destroys soft tissue of the nose and palate. Bone is not invaded.

III Sporozoa

Sporozoans are **intracellular protozoans** with **both sexual and asexual stages,** sometimes in different hosts. Their **apical complex** structure enables cellular invasion (they are **also called Apicomplexa**). "High-yield" sporozoa are ***Cryptosporidium, Plasmodium,*** and ***Toxoplasma.***

A. ***CRYPTOSPORIDIUM PARVUM*** is frequently found **in U.S. surface waters.** Because it is **not killed by water chlorination,** it is removed in treatment of drinking water by flocculation or filtration. It is usually acquired by swimming in or drinking untreated water. Municipal pools may get contaminated.
   1. **Cryptosporidiosis** is a **self-limiting diarrhea in healthy people. In IC patients, it causes chronic diarrhea with abdominal pain, fever, and anorexia,** resulting in weight loss and often death.
   2. **Lab ID** is via **microscopy** (finding **acid-fast oocysts** in feces or intestinal biopsies) or **fecal antigen tests.**

B. ***PLASMODIUM SPECIES*** **are sporozoans** that cause **malaria,** the most common *fatal* infectious disease in the world. Most malaria is acquired in the tropics or subtropics. Rare "airport malaria" is acquired from infected mosquitoes traveling on airplanes. Disease may occur many months after leaving the area.
   1. **Epidemiology. Four species of *Plasmodium*,** all carried by ***Anopheles*** **mosquitoes,** cause malaria. A general life cycle is shown in Figure 31-1 and outlined below.
      a. ***Anopheles*** **mosquitoes** inject **sporozoites** that **infect only liver parenchymal cells** and develop into hepatic/liver schizonts that contain thousands of daughter cells called *merozoites.*
      b. **In *P. ovale*** and ***P. vivax*** **infections only, liver schizonts persist** and remain quiescent for years (53 years is the record!). Primaquine is required to kill these persistent hepatic schizonts (formerly called *hypnozoites*). If primaquine is not used, **a recurrence of symptoms** (specifically called **relapse**) **comes from the persistent liver schizonts.** (M It's not "o v e r" if you treat ovale or vivax malaria *only* with chloroquine.)
      c. **Liver schizonts lyse 1 to 2 weeks later,** releasing **merozoites,** which only **infect red blood cells** (**not** reinfecting liver).

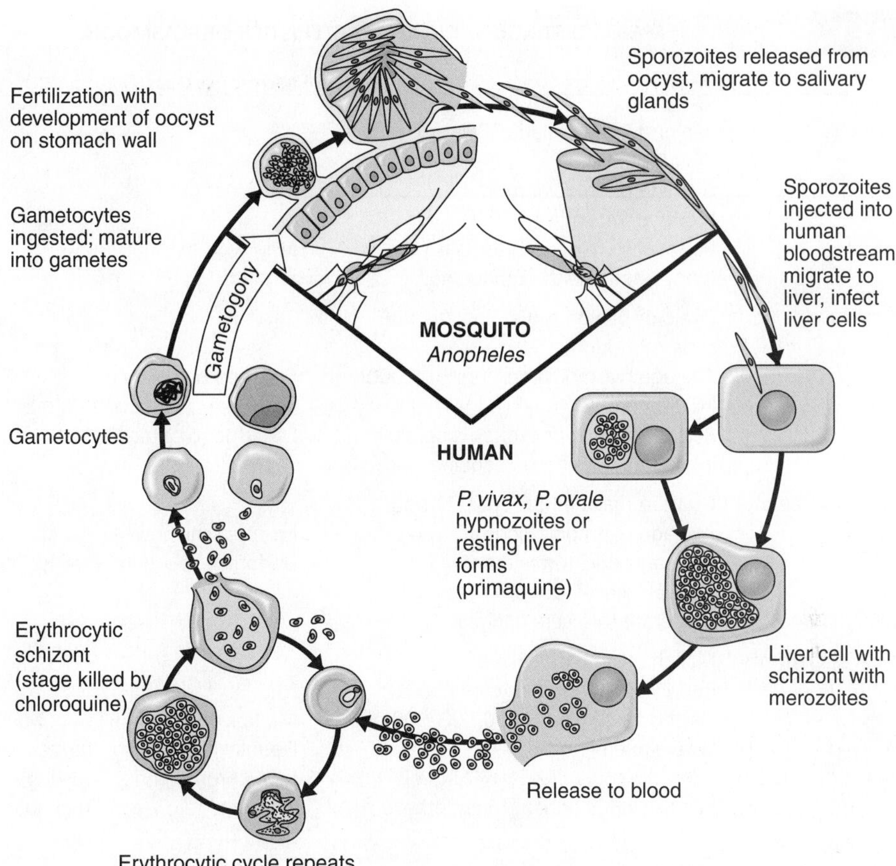

● **Figure 31-1** General life cycle of the *Plasmodium* species.

i. ***P. vivax*** and ***P. ovale*** infect only **reticulocytes** (<2% of the RBCs).
ii. ***P. malariae*** infects only **senescent red cells.**
iii. ***P. falciparum*** **infects all stages of RBCs,** resulting in prominent anemia and RBC surface changes, including knobs or adhesins, which cause infected cells to adhere to the capillary endothelium with decreased microcirculation, disseminated intravascular coagulation (DIC), and cerebral malaria. Duffy antigen-negative RBCs are resistant to merozoite invasion.

d. **RBC** stages. **Trophozoites** (ring stages) develop **into schizonts with more merozoites.** Synchronized release of large numbers of erythrocytic merozoites (called *sporulation*), toxic hemoglobin breakdown products, and triggered cytokines results in the typical malarial fever paroxysm. Eventually, some infected RBCs produce and release **gametocytes, making the person infectious to a mosquito.**
e. The **sexual cycle** begins with ingestion of the gametocytes by an *Anopheles* mosquito, followed by fertilization. The entire cycle is shown in Figure 31-1.

2. **Malaria symptoms** are species-dependent (Table 31-1). **Malaria always starts with an influenzalike prodrome, followed by periodic paroxysms, which can be summarized as cold, hot, wet.** Paroxysms start with shaking chills, a pale cyanotic appearance, and increasing fever, followed by a flushed, hot stage (up to 102 to 107°F) with agitation, disorientation, severe frontal headache, and limbic pain.

TABLE 31-1 MAJOR DISTINGUISHING CHARACTERISTICS OF PLASMODIA

Species	Disease	Microscopic Clues
Plasmodium vivax	Benign tertian malaria RBC cycle 44–48 hours	
	Average plasmodia per mm^3 = 20,000	Schüffner granules
	Moderate anemia	
	Relapse from liver schizonts is possible if not treated with primaquine.	May have two ring forms.
Plasmodium ovale	Ovale or benign tertian malaria RBC cycle 48 hours	
	Average plasmodia per mm^3 = 9,000	Enlarged oval red cells
	Mild anemia	
	Relapse from hypnozoites is possible if not treated with primaquine.	Schüffner granules
Plasmodium malariae	Quartan malaria RBC cycle 72 hours	
	Average plasmodia per mm^3 = 6,000	Rosette schizonts
	Persistent RBC forms may lead to recrudescence* up to 30 years later.	Uniformly round infected RBCs
	Moderate to severe anemia	
Plasmodium falciparum	Malignant tertian malaria	
	RBC cycle 48 hours or continuous	
	Plasmodia per mm^3 = 50,000–500,000	Multiply infected RBCs (1–3 per cell)
	Severe anemia	Double "signet ring" forms (two chromatin dots per ring)
	Infected RBCs adhere to endothelium.	Schizonts rarely seen in peripheral blood due to RBC adherence
	CNS and all organ involvement	Crescent-shaped gametes
	Medical emergency	

*Ⓜ Recrudescence is a reoccurrence of symptoms from persistent red cell forms and only occurs with *Plasmodium malariae*.

Sweating occurs as the fever defervesces, followed by severe exhaustion and deep sleep. Patients are often asymptomatic when they wake up. Falciparum malaria is most likely fatal, due to the nodular membrane adhesins triggered on the infected RBCs, which cause them to adhere to the endothelium of capillaries and postcapillaries in the brain and all other organs. (This is a gross oversimplification of a very complex organism's virulence and the immune system response that, in heavy parasitemia, also causes damage.) **Two to three million deaths** occur a year, mainly in African **children** and **pregnant women.**

3. **Lab ID** of ***Plasmodium*** is by **microscopic examination** (see Table 31-1). The ability to maintain the entire life cycle in the lab, the sequencing of the genome, and cell hybridomas producing specific antibodies have led to **new antigen detection methods** and the ability to test new drugs.
4. **Prophylaxis and treatment.** (For a more extensive discussion of the drugs, see *High-Yield Pharmacology*.) Total drug eradication of malaria would require killing liver schizonts, erythrocytic schizonts, and gametocytes. Unfortunately, **most drugs hit only one stage.** (The exception is artemisinin, which kills erythrocytic schizonts and prevents gametocyte development, so a person does not become infectious.) Erythrocytic stages are killed by chloroquine, quinine, the antifolates/sulfonamides

(plasmodia make their own folates), and artemisinins. Thus, **malarial prophylaxis** (drugs killing erythrocytic forms) **must be continued for 4 weeks after leaving a malaria region** to allow **all liver schizonts to convert to RBC forms and be killed.** If *P. vivax* or *P. ovale* infection occurs, treatment with **primaquine must be added to prevent relapse.** (It is the only drug that kills liver schizonts.) Chloroquine-resistance *P. falciparum* is widespread in Africa and Asia and is increasing in other places; the organism is developing resistance to other drugs. Prophylaxis now depends on where the traveler is going, the traveler's genetics, and the traveler's state of health. **Treatment of malaria in a nonimmune (i.e., nonindigenous) person is a medical emergency,** with initial hospitalization recommended. Pregnant or IC patients have increased risks.

C. *TOXOPLASMA GONDII* is a **sporozoan** that causes **toxoplasmosis,** a major problem in AIDS patients, due to its widespread nature. It is found in almost all animals.

1. **Epidemiology** (Figure 31-2). **Sexual stages** occur **only in** the gut of **cats** (which are the **definitive hosts**). The **feline-excreted oocysts** (a thick-walled, hardy stage) **mature in 1 to several days** in the environment, where they **may remain viable for years in soil.** All **vertebrates** serve as **intermediate hosts.** Rodents that eat cat feces contaminate fields, infecting cattle, sheep, pigs, and other animals. **Human exposures** occur **through ingestion of undercooked meat with tissue cysts,** contamination with **cat feces oocysts,** or **transplacental transfer.** In the

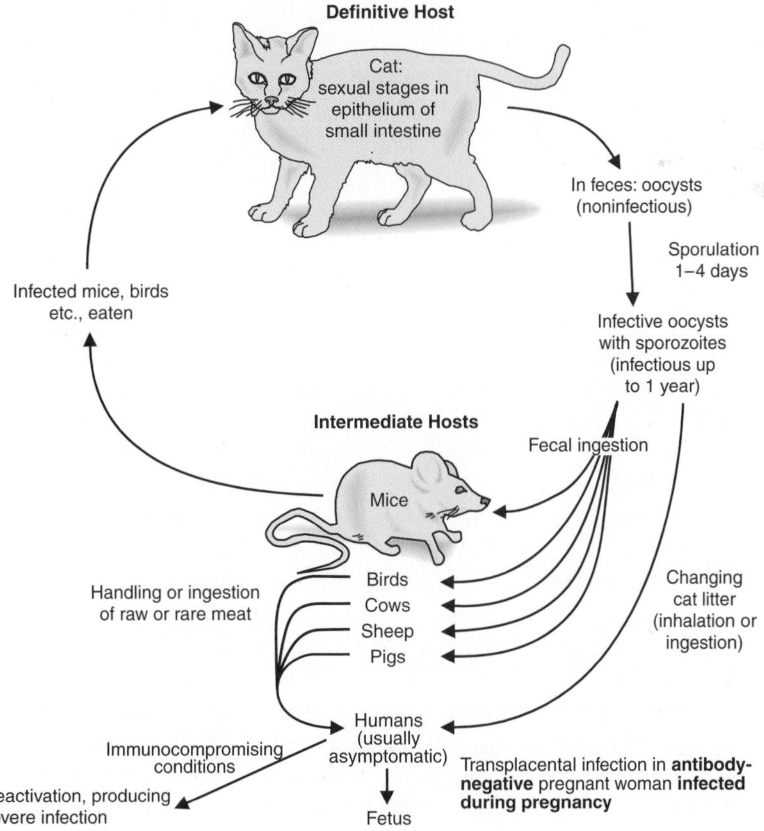

● **Figure 31-2** Life cycle of *Toxoplasma gondii.* (Note human involvement at bottom of figure.)

initial infection, *Toxoplasma* **ultimately creates tissue cysts** that may remain viable for years, **producing long-lasting antibody titers** and **potential for later relapse** under immunocompromising conditions. **Antibodies** in women infected **prior to pregnancy prevent transplacental transfer.** In antibody-negative women infected during pregnancy, the baby may be infected.

2. **Human toxoplasmosis**
   a. **Healthy adults** usually have benign asymptomatic or mononucleosis-like infections, resulting in **antibodies** and often **persistent tissue cysts.**
   b. In antibody-positive AIDS patients (i.e., they have tissue cysts) with **CD4+ <100 per ml (with no prophylaxis),** toxoplasmosis may **reactivate,** leading to retinochoroiditis and focal CNS lesions.
   c. **Fetal infections** range from **asymptomatic (infected late)** to severe **(infected early: encephalomyelitis with calcifications, retinochoroiditis, and either hydrocephalus or microcephaly). Asymptomatic but infected neonates** (usually infected in the third trimester) **may develop retinochoroiditis if not** identified and **treated.** In utero damage is not reversible, but further damage can be prevented by treatment.
3. **Lab ID** is made via fleeting **high IgM titer, increasing IgG levels, PCR, or antigen tests in IC patients.** PCR on amnionic fluid is safe and sensitive.

Chapter 32

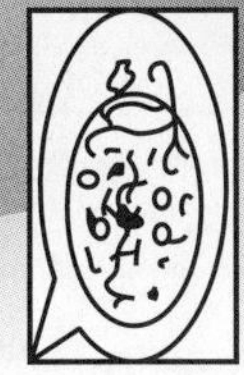

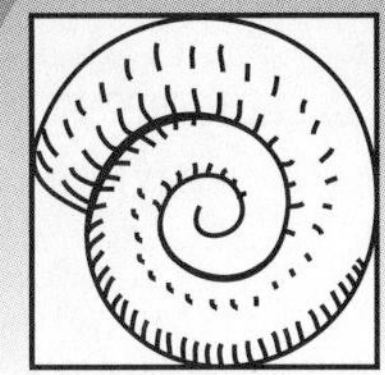

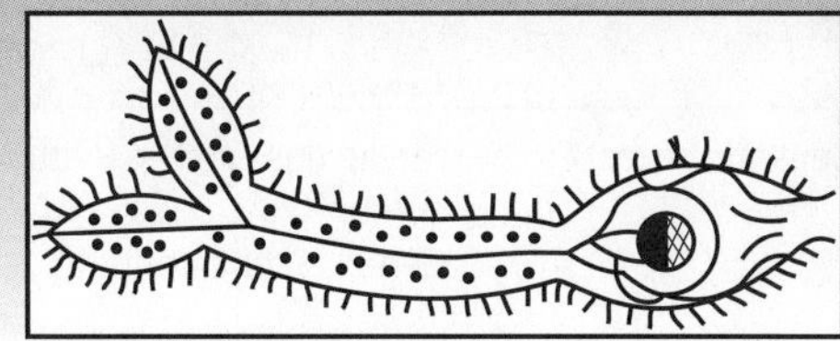

Trematodes (Flukes)

Fasciolopsis, Clonorchis, Fasciola, Paragonimus, Schistosoma

I Trematodes/Flukes*

Trematodes or flukes are **highly organized, nonsegmented flatworms** with an incomplete gastrointestinal (GI) tract. They all have complex life cycles involving **snails as intermediate hosts** and **water or aquatic animal or plant transmission to humans** or other definitive hosts. Characteristics of the **two types of flukes (separate-sexed schistosomes** and the **hermaphroditic)** are outlined in Table 32-1.

II Intestinal Flukes

Fasciolopsis buski (the giant intestinal fluke) is found in Asia and India. Human infection occurs when **cysts on water plants (bamboo sprouts, water chestnuts) are ingested. Diarrhea** is a symptom of infection.

III Liver Flukes

A. ***CLONORCHIS SINENSIS*** (the Chinese liver fluke) is transmitted by **raw or undercooked freshwater fish.** In infected humans, the adult worms localize in distal parts of the biliary tree, **sometimes living more than 30 years.**

B. ***FASCIOLA HEPATICA*** (the sheep liver fluke) is cosmopolitan in sheep-raising areas. (Ⓜ Add "s," making it *shepatica,* to remind you of sheep.) Transmission is by **ingestion** of metacercariae **(larvae) on watercress or other aquatic plants.**

*Ⓜ If you have trouble remembering that the scientific name for flukes is Trematodes, think flukes (flat and fleshy), flutter, tremor, Trematodes.

TABLE 32-1 TWO MAJOR TYPES OF FLUKES (TREMATODES): CHARACTERISTICS, TRANSMISSION, AND DIAGNOSIS

Type	Transmission	Diagnostic Stage
Schistosomes—Separate sexes	Larvae (in water) penetrate human skin. Genus: *Schistosoma*	Eggs with spines (no operculum)
Hermaphrodites	Encysted larvae are eaten in raw or undercooked food. Genus (infected food): *Fasciola* (watercress) *Fasciolopsis* (water chestnuts, water plants) *Paragonimus* (crab) *Clonorchis* (fish)	Operculated eggs ("hinged lid")

IV Lung Fluke

Paragonimus westermani is found in India, Asia, and Africa. Human infection starts with **ingestion of infected shellfish. Adult flukes in the lungs cause fever, cough, and hemoptysis. Eggs** are expelled **in sputum and feces.**

V Blood Flukes (Schistosomes)

A. SHARED SCHISTOSOMAL CHARACTERISTICS

1. **The male** is **flat** but **folds in ventrally to hold the female** (longer, slender, and rounder) in permanent copulation.
2. **Schistosomal larvae** (found in **water**) **penetrate human skin,** causing early **dermatitis. Adult flukes mature in abdominal veins.**
3. **Schistosomal egg spines** (see Figure 30-1) play an important role in disease and spread. Granulomas may form around the eggs. The **physical irritation from the spine and egg enzymes erodes the venules and underlying tissues, liberating the eggs into the intestine or bladder lumen.** Schistosomes may live many years in humans.
4. **Schistosomiasis** is one of the most common infectious causes of death in the world. About 150 million people are infected; there are approximately 1 million deaths per year, primarily among heavily infected native people.

B. *SCHISTOSOMA MANSONI* (seen in Africa, the Middle East, South America, and the Caribbean) causes **intestinal schistosomiasis.** Adult flukes mature in the mesenteric venules. Patients may be asymptomatic or have diarrhea, abdominal pain, and hepatosplenomegaly. Eggs with lateral spines in stool are diagnostic. Ⓜ *S. mansoni* eggs are the ones that sort of look like a cross section of the torso of a man (mansoni).

C. *SCHISTOSOMA JAPONICUM* (found in the Far East in rice paddies) also causes **intestinal schistosomiasis,** but eggs are more frequently carried in the circulation to distant sites (often to the choroid plexus and venules around the spinal cord).

D. ***SCHISTOSOMA HAEMATOBIUM*** (seen in Africa and the Middle East) causes **vesicular schistosomiasis. Adult flukes reproduce in bladder veins.** Eggs (which have a terminal spine) are trapped in the bladder venules, where they cause granulomas and erosion, resulting in **urgency, frequency, dysuria, and hematuria.** Bladder carcinoma in Egypt has a high association with chronic schistosomiasis.

E. ***SCHISTOSOMES*** that normally infect aquatic birds and animals may cause schistosomal dermatitis **(swimmer's itch)** in humans. These organisms die after penetrating the human skin.

 Treatment of Fluke Infection

Praziquantel is used to treat **all trematode infections**, **except swimmer's itch**, which is treated with calamine.

Chapter **33**

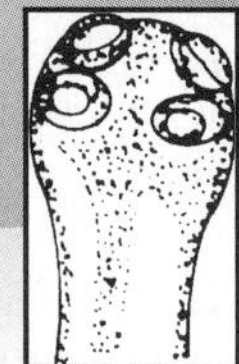

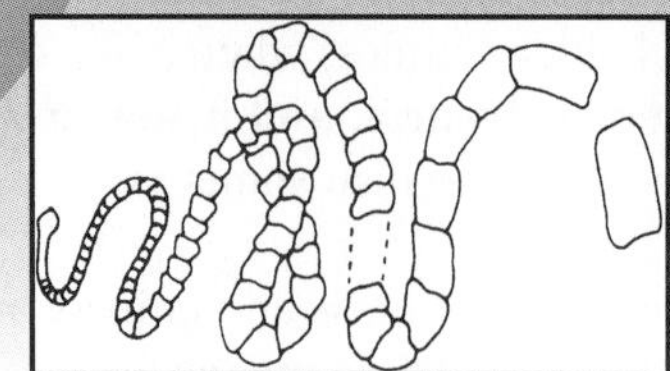

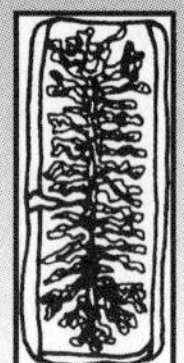

Cestodes (Tapeworms)

Taenia, Diphyllobothrium, Echinococcus

I Cestode Basics

A. CESTODES are **segmented flatworms. Adults** have **three types of segments:** the **scolex** (head); the **neck region,** which produces numerous proglottids; and the **strobila,** the linearly arranged proglottids that make up the major part of the worm. Each proglottid contains both a female and a male genital tract. Mature distal proglottids produce fertilized eggs.

B. LIFE CYCLE. Cestodes have complex life cycles involving **two or more hosts.** Be they human or not, **intermediate hosts (IHs)** are the **hosts ingesting eggs (or early larvae,** if there is an early and late larval form). **IHs develop severe disease,** because of the invasiveness of the larvae, which bore out of the gastrointestinal (GI) tract into the lymphatic and vascular system and then wander through tissues. During this acute wandering, eosinophilia is present. **Symptoms vary, depending on where the cysticerci** (or equivalent structures) **develop.** Fortunately, humans cannot serve as intermediate host for some cestodes, and with sanitation and good kitchen techniques, we are less frequent hosts for others. **Definitive hosts (DHs) ingest larval forms in undercooked fish or meat** and develop **adult tapeworms in the GI** tract. Unless there is a heavy worm burden, the tapeworms may be **asymptomatic** or simply cause **vague abdominal discomfort, vomiting, and weight loss.**

II Intestinal Tapeworms (Humans as Definitive Host)

A. *TAENIA SOLIUM** (**pork tapeworm,** rare in the United States, except in immigrants) and ***T. saginata*** (**beef tapeworm,** in ~1% of beef sold in the United States). When larval forms (cysticerci) in undercooked pork, game, or beef are ingested by humans, **adult**

* Ⓜ *T. solium* = pork tapeworm (Note: *Trichinella* is the pork roundworm.)
T. saginata = beef tapeworm (Think astrological symbol of Sagittarius.)

tapeworms (up to 5 m long) develop in the intestine. Thick-walled, spherical, dark brown eggs in the feces or proglottid segments are diagnostic. The sticky tape technique used for pinworms increases diagnostic success for adult tapeworms.

B. *DIPHYLLOBOTHRIUM LATUM*[†] (The broad fish tapeworm)

1. ***D. latum* proglottids are wider than they are long** (thus the nickname). Adult tapeworms can grow **up to 10 m.** This tapeworm has been found in fish from northern areas, such as Alaska (including salmon) and Canada, as well as in those from Florida, California, and around the world. (There is a potential increase with popularity of sushi. Freezing at –10°F for 48 hours kills larval forms.)
2. **Infection.** Ingestion of infected raw, undercooked, salted, or pickled fish can lead to **largely asymptomatic infections** of the small intestine. ***D. latum* competes with the host for vitamin B_{12}.** If *D. latum* attaches **high in the jejunum, macrocytic hyperchromic anemia** or neurologic manifestations may occur through **B_{12} deficiency.**

III Tissue Cestodes (Humans as Intermediate Host)

A. CYSTICERCOSIS. If **humans ingest embryonated eggs of *Taenia solium,*** often from their own intestinal tapeworm, **the hatched larvae exit the intestinal wall to the bloodstream and may infect any tissue,** most seriously the eye (with visual loss) or the brain. Symptoms of *acute* central nervous system (CNS) invasion include **headache, fever, and eosinophilia.** Later symptoms depend on where the multiple cysticerci (cysts) develop in the brain. In many endemic areas, cysticercosis is the major cause of epilepsy. Symptoms and a vigorous inflammatory response with eosinophilia are triggered by death of the cysticerci. Diagnostic tools include biopsy of subcutaneous nodules, imaging (the kidney bean–shaped cysticerci calcify), and cerebrospinal fluid (CSF) and serum tests for immunoglobulins.

B. HYDATID CYST DISEASE. Accidental human **ingestion of the embryonated eggs of *Echinococcus granulosis* (dog, wolf tapeworm)** from canine fecal contamination (often of fur) results **in the growth of the embryo into a hydatid cyst.** This space-filling structure, which, over 10 years, can grow up to 20 cm in diameter, is most commonly located in the lung or liver. The hydatid cyst contains infectious protoscoleces (*scoleces* is plural of *scolex*), hydatid sand, and an anaphylatoxin. Treatment is with albendazole, followed by aspiration of the cyst contents and subsequent injection of scolecidal fluid. This treatment is replacing very careful surgical removal. Deworming of sheep dogs with praziquantel every 6 weeks and not letting them eat uncooked viscera of slaughtered sheep has reduced the incidence.

C. ALVEOLAR HYDATID CYST (from *E. multilocularis*) is similar to hydatid cyst disease, except that multiple smaller cysts bud out, **producing a structure that resembles alveoli.**

[†] Ⓜ Diphyllobothrium (Think about two fishermen who don't bother to cook their fish: Di-fish-no-botherum.)

Chapter 34

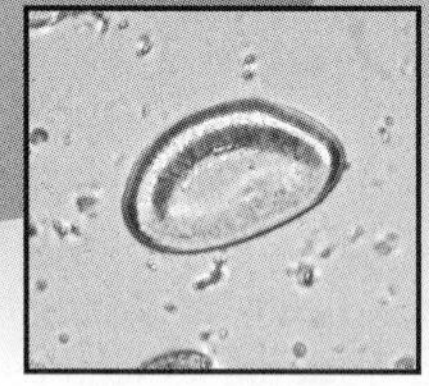

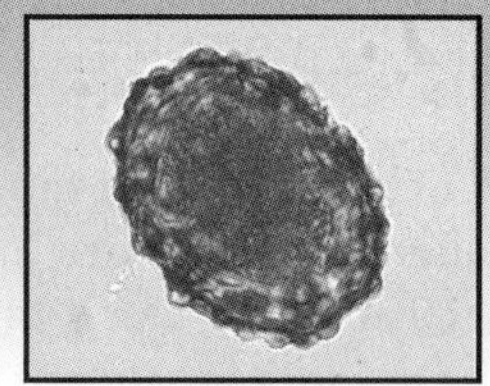

Nematodes (Roundworms)

Enterobius, Ascaris, Trichuris, Trichinella, Strongyloides, Ancylostoma, Necator, Wuchereria, Brugia, Loa loa, Onchocerca, Dracunculus

I Nematode Basics

A. DEFINITION. Nematodes are roundworms. They have **unsegmented bodies** and **separate sexes** (males are usually smaller), and they produce **large numbers of eggs** (up to 1 million per female in some species). **Larvae** hatch from eggs, **develop through a series of molts**—some requiring warm, moist soil; others, in a host. The third, and final, stage larva (filariform) is often a very invasive form. Where skin penetration starts the infection, the filariform larvae are the infectious form, even when mosquitoes help deposit it. Filariform larvae develop into the adults.

B. INFECTION. Three major patterns of infection predominate. The first pattern is the **ingestion of eggs that cause intestinal worms. The second is the penetration of skin by the filariform larvae from the soil, which leads to intestinal infections. The third is by mosquito bite that deposits the filariform larvae for the filarial tissue worms. In all nematode infections, if there is migration in vasculature, it produces eosinophilia. Severity of symptoms depends on worm burden, previous sensitization, and nutritional status of the infected individual.** Current or previous infection does not prevent reinfection, which is common in endemic areas of poor hygiene or infected mosquitoes. Some nematodes are long-lived. *Strongyloides* infections, the eggs of which develop into the filariform larvae while still in the intestine, can reinvade the tissues in increasing numbers (autoinfection) and significantly increase the worm burden, even with no further exposure in the environment.

C. OVERVIEW OF INTESTINAL NEMATODES. Some forms are ingested (Ⓜ *Enterobius Ascaris, Trichuris, Trichinella*—EAT^2), whereas others penetrate skin (Ⓜ *Strongyloides, Ancylostoma, Necator*—S[k]AN). Intestinal roundworms usually mature in the small intestine, where most adults attach by anterior oral hooks or cutting plates. These nematodes cause disease by blood loss, irritation, and allergy. Their invasiveness spreads the roundworm larvae, which may result in secondary bacterial infections.

D. OVERVIEW OF TISSUE NEMATODES. Tissue nematodes are **spread by mosquitoes (*Wuchereria* and *Brugia*)** or **biting flies (*Loa loa* and *Onchocerca*).** The exception is ***Dracunculus*, which is ingested in contaminated drinking water.** Adult tissue nematodes produce symptoms of irritation or inflammation, due to nematode secretions or components in the lymphatics. This leads to lymphangitis and, ultimately, elephantiasis of genitalia or limbs.

E. Ⓜ MNEMONIC: NEMA²T³ODES
Necator (U.S. hookworms)
Enterobius (pinworms)
Mosquito-borne: *Wuchereria* and *Brugia*
A²—*Ascaris* and *Ancylostoma*
T³—*Trichuris*, *Trichinella*, and *Toxocara*
Onchocerca (river blindness)
Dracunculus (Guinea worm; nearly eradicated)
Eye worm (*Loa loa*)
Strongyloides (threadworm)

II Intestinal Nematodes

A. *ASCARIS LUMBRICOIDES.* This large nematode (adults up to 0.5 to 35 cm) is the **most common roundworm—as well as the most common worm—in the world** (approximately 1.3 billion people are infected). In areas of poor sanitation, nearly everyone is infected, often with heavy worm burdens.

1. **Life cycle. *Ascaris* eggs are ingested. Larvae hatch, exit the duodenal wall,** and **migrate through the vasculature** into the **liver, heart,** and **lungs. In the lungs,** they **burst out into the alveoli, causing** a pneumonitis. **As with other worm migrations through the vasculature, eosinophils and IgE levels may be elevated.** Molting, migrating, and being carried by cough, **the larvae eventually reach the esophagus and enter the gastrointestinal (GI) tract again.** There they mature in the small intestines, where the females lay eggs.
2. **Disease: Ascariasis. Adult ascarids do not attach;** rather, they **maintain their position in the intestines by motility.** Infected people with low worm burdens are often asymptomatic. In children with heavy worm burdens, the worms may tangle and produce an intestinal blockage. **Increased worm migration is seen with fever, anesthesia,** or **antibiotic use** and **may result in blockage of bile ducts or appendix or in migration up the esophagus or into the gallbladder.**

B. *ENTEROBIUS VERMICULARIS* (pinworm) is the **most common roundworm infection in the United States.**

1. **Life cycle and infection. *Enterobius*** infection begins with **ingestion of eggs in household dust** from contaminated clothing or bedding. Adult female pinworms (up to 0.5 in. long) migrate out to perianal and vaginal areas and lay eggs, causing intense perianal itching. **Autoinfection occurs from perianal scratching and then putting fingers in the mouth.**

2. **Diagnosis.** A tongue depressor with scotch tape **(sticky side out)** has been used for a perianal sampler. A plastic paddle or slide (in a tube) with one sticky surface is now available for obtaining samples. (This is usually done at home.)

C. **HOOKWORMS.** ***Necator americanus*** (New World hookworms, endemic in the southeastern United States) and ***Ancylostoma duodenale*** (Old World hookworms) thrive in warm, humid climates.

1. **Life cycle and disease. Eggs in feces deposited into soil develop into filariform larvae, which attach between the toes of bare feet** and penetrate skin. They migrate through the bloodstream into alveoli, causing a **pneumonitis.** They then migrate up the trachea or are coughed up and swallowed, reentering the GI tract and attaching to the small intestine. Repeated infections lead to **allergy** at entry points **("ground itch")** and more severe pneumonitis. Infections are commonly chronic, leading to **weight loss, microcytic hypochromic anemia,** and fatigue, particularly if nutrition is poor.
2. **Prevention is by wearing shoes** and/or improving sanitation.

D. ***STRONGYLOIDES STERCORALIS*** **(threadworm)** is found in the soil of tropical and subtropical areas (e.g., southeastern United States) from human fecal contamination. Like hookworms, ***S. stercoralis*** **penetrates feet and migrates through the vasculature.** Unlike hookworms, **some larvae develop that reinvade the intestinal wall and begin an autoinfection** without leaving the human host to develop in soil. Thus, untreated people may be infected for years.

E. ***TRICHURIS TRICHIURA*** **(whipworm)** is found in the *tropics* in areas of **poor sanitation.** Eggs are ingested, larvae develop in the small intestine, and adults attach in the large intestine. **Trichuriasis** is **frequently asymptomatic.** In heavier infections, **it causes abdominal pain with mucoid or bloody diarrhea, tenesmus (sometimes resulting in rectal prolapse), and weight loss.** It worsens poor nutrition.

F. ***TRICHINELLA SPIRALIS*** **(pork roundworm)** has both intestinal and tissue stages in humans.

1. **Intestinal stage.** Encysted larvae are ingested in **undercooked pork or wild game** (e.g., bear or deer) and develop into adults. **Early symptoms** are **nausea and abdominal pain,** with **vomiting and diarrhea, fever, and headache.**
2. **Tissue stage. Female worms burrow to lay eggs in submucosa** (so there are **no eggs in stools**). **Larvae hatch and migrate into the bloodstream, then out into striated muscle,** causing degenerative and inflammatory changes in the fibers. The **coiled larvae may calcify.** The **symptoms of the circulation and tissue stages** usually come **2 to 8 weeks after the GI symptoms** and include **high fever, eosinophilia (10% to 90%), myalgia, periorbital edema, conjunctival and subungual (splinter) hemorrhages, and urticarial rash.**

III Tissue Nematodes: Filarial Worms

All filarial worms are transmitted to humans via an arthropod bite. None is significant in the United States. **Only the scientific name, common name, and vector are "high-yield" information.** The adults are **all threadlike worms that live in lymphatics and subcutaneous tissues** for years. **Microfilariae (their live-borne larvae) migrate through the vasculature and subcutaneous region and are picked up through mosquitoes.**

A. ***ONCHOCERCA VOLVULUS*** (spread by the ***Simulium*** **blackfly**) causes **river blindness,** one of the World Health Organization's six most harmful infectious diseases. **Subcutaneous nodules around adult nematodes** are prominent early. Sensitization of the human host to the filariae, which then migrate through the body, leads to allergic-toxic reaction, including **abnormal skin and subcutaneous tissue reactions and eye reactions, which leads to blindness.**

B. ***LOA LOA*** **(eye worm)** is found in Africa and has a somewhat similar progression to that of *Onchocerca*. It is spread by the *Chrysops* biting fly.

C. ***WUCHERERIA BANCROFTI*** is a **tropical filarial worm** spread by a variety of **mosquitoes.** The microfilariae are in the blood (causing **fever, chills,** and **eosinophilia**). **Granulomas from adult worms in lymphatics lead to elephantiasis.**

D. ***BRUGIA MALAYI*** is a **Far East filarial,** transmitted by a variety of **mosquitoes.** It produces a **disease similar to *Wuchereria*** infection.

IV Dracunculus

When ***Dracunculus medinensis*** **(Guinea worm) larvae in tiny aquatic copepods are ingested** and become **mature, painful subcutaneous lesions (often on a foot) develop. Adult *Dracunculus* (up to 1 m) may be removed either by a very careful, slow pull (rolled on a stick and eased by antiparasitic drugs) or surgically.** In some areas, filtering drinking water through clean t-shirt material to remove the crustaceans has nearly eradicated *Dracunculus* infection in humans.

V Animal Roundworms

Animal roundworms may infect humans; however, humans are dead-end hosts.

A. **DOG OR CAT ASCARIDS** penetrate human skin but not the vasculature, so they "creep" around in cutaneous and subcutaneous tissues, causing **cutaneous larva migrans.**

B. **DOG AND CAT *TOXOCARA*** nematodes (***T. canis*** and ***T. cati***) are usually asymptomatic in humans. They are associated with pica.

VI Treatment of Nematode Infection

Albendazole is now the drug of choice for all roundworms, except for *Strongyloides*.

Part VI
Infectious Diseases

"Paper" exam cases, such as those on the USMLE Step 1 exam, are different from real cases encountered in the practice of medicine. Paper cases usually fall into one of two categories: (1) a common disease with fairly clear presentation and either a major known causative agent (CA) or easily distinguished multiple CAs; or (2) an uncommon but serious disease with a distinctive presentation and CA. For example, diphtheria is a serious and distinctive disease (with a toxin), so it is more likely to be featured in a paper case than is viral meningitis, which has many viral CAs.

Chapters 35 through 43 present infectious diseases as high-yield cases—brief case abstracts with no unnecessary details. The format is designed both for study and for self-testing. For self-testing, cover each page with a piece of paper and reveal one line at a time. The question icon (Q) indicates the transition from the case information to questions related to the case. The questions are based on the most likely test material. Stop at the A answer icon to formulate your answers. Then check your answers and move on to the next case.

Chapter **35**

Eye and Ear Infections

I Eye Infections and Conditions

A. A painful, hot, red swelling around an eyelash follicle Q Causative agent (CA)? Name of swelling? A
Staphylococcus aureus; stye (hordeolum)

B. Bilateral eyelid swelling, along with muscle pain and eosinophilia Q CA? A
Trichinella spiralis

C. Unilateral swelling around one eye in a South or Central American resident (usually from areas of poverty) or foreign workers in these areas; chronic infection may lead to enlarged, flaccid heart; associated with heart failure. Q CA? Transmission? What form would be found in the heart tissue? A
Trypanosoma cruzi; reduviid bite; amastigotes

D. Conjunctivitis (red eye)

1. First day of life; watery exudate and hyperemia Q CA? A
   No infectious agent—silver nitrate susceptibility
2. Neonate (1–4 days old) born at home without midwife with hyperpurulent exudate; Gram-negative (Gr–) bacterium Q CA? Seriousness? A
   Neisseria gonorrhoeae; it is rapidly destructive and causes irreversible damage.
3. Neonate (3–10 days) with purulent conjunctivitis (may have been born in hospital with the best of medical care); inclusion bodies seen in cells Q CA? Which serotypes? What happens if not treated? A
   Chlamydia trachomatis; serotypes D–K; neonate will develop pneumonia.
4. Purulent postneonatal conjunctivitis Q Gram+ CA? Gram– CA? A
   Streptococcus pneumoniae or *Haemophilus influenzae* (formerly *H. aegyptius*)
5. Conjunctivitis in children, with watery exudate ± sore throat; often associated with swimming Q CA? Why swimming association? A
   Adenovirus, a naked capsid virus, so it is somewhat resistant to chlorination

E. Follicular conjunctivitis, leading to in-turned eyelashes, corneal scarring, and loss of vision; found in deserts of southwestern United States, especially in Native Americans Q CA? Serotypes? Disease? A
Chlamydia trachomatis, serotypes A, B, Ba, C; trachoma

F. Eye pain and ulcers from (1) wearing extended-wear contacts too long or (2) being in a coma Q CA? A
Pseudomonas aeruginosa

G. Chorioretinitis in neonate or AIDS patient (Sometimes lesions are described as looking like catsup and mustard.) Q CA? A
Toxoplasma gondii is most common.

H. Eye pain in soft contact wearer who has made homemade saline Q CA? A
Acanthamoeba (free-living ameba)

II Ear Infections

A. Acute otitis media with effusion Q Most common CAs? (Hint: three bacteria and three viruses) A
Streptococcus pneumoniae is most common, but also nontypeable *Haemophilus influenzae* or *Moraxella catarrhalis* (Gram + diplococci) and other bacteria. Viruses include respiratory syncytial virus (RSV), adenovirus, or influenza.

B. Otitis malignant externa in diabetics Q CA? A
Pseudomonas aeruginosa

Chapter 36

Respiratory Tract Infections

I Upper Respiratory Tract Infections and Manifestations

A. Exudative, erythematous pharyngitis; fever of ≥101°F (38.3°C); cervical lymphadenitis; positive rapid-antigen screen; backup cultures show beta-hemolytic, Gram-positive (Gram+) cocci inhibited by A disk. Q Causative agent (CA)? What is in the A disk? A
Streptococcus pyogenes (group A strep); bacitracin

B. Teen or young adult with severe fatigue, pharyngitis, cervical lymphadenopathy, spleen enlargement, abnormal white blood cells (WBCs). Group A strep rapid-antigen screen and backup cultures are negative. Q CA? What types of WBCs are the abnormal WBCs? With what do heterophile antibodies cross-react? A
Epstein-Barr virus. Downey type II cells are reactive T cells. Heterophile antibodies cross-react with animal red blood cells (RBCs), not with any viral antigens.

C. Pharyngitis with fever <101°F (38.3°C); Group A strep rapid-antigen screen and backup cultures are negative. The virus causing it is a naked double-stranded (ds) DNA virus Q CA? Unusual exterior feature of this group of viruses? A
Adenoviruses; fibers

D. Child with sore throat, fever, and gray-white papulovesicular lesions on soft palate, anterior pillars of the tonsillar fauces, uvula, and tonsil; usually without gingivitis. Q CA? Disease name? Replicative intermediate? A
Coxsackie A; herpangina. Replicative intermediate is (–) RNA.

E. Unvaccinated child, recently arrived from former Soviet Union, presenting with sore throat with dirty pseudomembrane and heart irregularity. Q CA? Disease? How does the major virulence factor work? A
Corynebacterium diphtheriae; diphtheria. A-B component exotoxin is produced and circulates (bacterium is not invasive), toxin ADP-ribosylates elongation factor 2 (EF-2), turning off protein synthesis. Mainly affected are epithelial, nerve, and heart cells.

F. Common cold Q CA prominent in hotter weather into fall? CA in winter or spring? A
Rhinovirus (primarily hot weather/fall); coronavirus (winter/spring)

G. Sinus pain and inflammation. Q CAs? A
Most commonly *Streptococcus pneumoniae;* less commonly *Moraxella catarrhalis* (Gram– and more drug resistant)

H. Paranasal swelling and hemorrhagic exudate in eyes or nose in a ketoacidotic or leukemic patient; mental lethargy Q CAs? What would you expect to see in the tissues? A
Rhizopus, Mucor, or *Absidia.* All three are nonseptate, extremely rapid growing, filamentous fungi with branching at nearly right angles.

I. Fever, chills, upper respiratory symptoms, severe arthralgias, and myalgias between November and March in the northern temperate zones. Leads to secondary pneumonias, particularly in the elderly. Q CA? What explains major changes in the virus leading to pandemics? A
Influenza virus; gene reassortment (genetic shift)

J. Starts with coldlike symptoms, progresses to a repetitive cough, ending in an inspiratory whoop and often vomiting. Q CA? Mechanism of the major virulence factor? A
Bordetella pertussis; pertussis toxin ADP-ribosylates G_i.

K. Viral look-alike to pertussis Q CA? A
Adenovirus

L. Croup; positive hemagglutination inhibition test Q CA? A
Parainfluenza virus

M. Epiglottitis, usually in child <3 years old; male > female; presents with drooling; needs to lean forward to breathe. Q CA? What part of well baby care is child most likely missing? A
Haemophilus influenzae; vaccine of polyribitol capsular material linked to protein

II Bronchitis and Bronchiolitis

A. Bronchiolitis in infant or child younger than 2 years old Q CA? A
Adenovirus

B. Bronchiolitis in child 2 to 5 years old Q CA if hemagglutinin negative? If hemagglutinin positive? A
Respiratory syncytial virus (hemagglutinin negative); parainfluenza virus (hemagglutinin positive)

C. Bronchitis in asthmatic patient Q CAs? A
Most commonly viral, but also *Streptococcus pneumoniae, Haemophilus influenzae, Moraxella catarrhalis.* (USMLE exam questions would provide specifics if bacterial.)

III Pneumonia

(Note: Age is an important predictor of the most common CA, which is listed first. If you are asked to provide other CAs on the exam, you will be given specific CA characteristics.)

A. Pneumonia in nonfebrile neonate who presents with a staccato cough, no inspiratory whoop, ± conjunctivitis, and respiratory distress; x-ray pattern infiltrative, hyperinflation of lungs. Q CAs? What is the infective form? A

Most commonly *Chlamydiae trachomatis,* which is acquired at birth but may not show clinically until 3 months. The transmitted form is the elementary body. There are many other CAs (but none as distinctive): cytomegalovirus (CMV), rubella, herpes simplex virus (HSV), *Streptococcus agalactiae,* and *Streptococcus pneumoniae.*

B. Pneumonia in child 1 to 3 months old. Q CAs? A
Chlamydia trachomatis (as above), RSV or other respiratory virus, or *Bordetella* (very severe in young)

C. Pneumonia in a child 3 months to 5 years old Q CAs? A
Usually viral (RSV or other), but also *Streptococcus pneumoniae, Chlamydia, Haemophilus influenzae, Mycoplasma.*

D. Pneumonia in a child 5 to 18 years old; sore throat leading to nonproductive persistent hacking cough; no growth on blood agar; mixed flora on induced sputum. Q CA? What antibiotic would not be effective? What is required growth factor? A
Mycoplasma (the most common CA in this age group; incidence decreases with increasing age). Beta-lactam antibiotics are not effective; sterols are growth factor.

E. In all age groups, the most common causative agent of lobar pneumonia, presenting with fever and productive cough with blood–tinged sputum. Q CA? What is this organism's catalase reaction? A
Streptococcus pneumoniae (also most common in alcoholics, people >60 years, or smokers). It is catalase negative.

F. Pneumonia in adult with upper respiratory symptoms, diffuse infiltrates, and hypoxemia (November to March) Q CA? A
Viral pneumonia, possibly influenza

G. Other causes of bacterial pneumonia Q CAs? Distinguishing features of each? (Write out answer!) A
Haemophilus influenzae (Gram–, may be lobar); *Chlamydia pneumoniae* (non-Gram-staining, usually mild disease); *Staphylococcus aureus* (Gram+, catalase positive, may be lobar); *Legionella* spp. (acquired from air conditioning; poorly Gram staining, atypical; requires charcoal yeast extract [CYE] medium); *Chlamydia psittaci* (after exposure to sick birds or their dried excretions)

H. Pneumonia in patient who has aspirated and now has foul-smelling sputum Q CA? A
Anaerobes

I. Pneumonia in an AIDS patient with a dry cough who is not receiving any prophylactic drugs Q CA if silver-staining cysts? If acid fast bacteria? If spherules? If tiny intracellular yeasts? A
Cysts: *Pneumocystis carinii*
Acid-fast bacteria: *Mycobacterium tuberculosis* or *Mycobacterium avium-intracellulare*
Spherules: *Coccidioides*
Tiny yeasts: *Histoplasma capsulatum*

J. Pneumonia in an AIDS patient with purulent sputum Q CA? A
Streptococcus pneumoniae

K. Pneumonia in teenage cystic fibrosis patient Q CA? A
Pseudomonas aeruginosa

L. Arthropod bite, leading to enlarged lymph nodes, high fever, conjunctivitis, and often pneumonia; found in southwestern United States Q CA? A
Yersinia pestis

M. Seriously ill older male smoker and heavy drinker has had rapid onset of pneumonia. Two of his bar buddies have the same acute onset of pneumonia, with major headache, mental confusion, and diarrhea but nothing seen on Gram stain of sputum. Organism does not grow on blood agar and would be seen on silver stain, but a direct fluorescent antibody test would be used for rapid ID of agent. Q CA? What medium is required? A
Mainly *Legionella pneumophila* and *Legionella micdadei;* poorly Gram staining, atypical pneumonia; CYE medium required (has cysteine and extra "iron")

N. Pneumonia with environmental associations:

1. Contact with sick birds or dried bird excretions Q CA? A
Chlamydia psittaci
2. Contact with dust enriched with bird or bat feces, in great river valleys of central United States Q CA? A
Histoplasma capsulatum
3. Contact with desert sand of southwestern United States Q CA? A
Coccidioides immitis

Chapter **37**

Nervous System Infections

I Meningitis

A. Viral meningitis (often called *aseptic meningitis*) has a tendency to be self-resolving and less acute than bacterial meningitis. **Q** Causative agents (CAs)? **A**
Enteroviruses (like coxsackieviruses and echoviruses), as well as mumps, polio, and some of the arboviruses

B. Purulent meningitis, rapid onset, all ages except neonates **Q** CA? Highest number of cases in what groups? **A**
Streptococcus pneumoniae is the most common causative agent of meningitis, with highest number of cases in the very young (except neonates) and in people >65 years of age.

C. Meningitis in neonates, particularly if there is a prolonged rupture of membranes and if Mom has had multiple sex partners **Q** Most common CA? What test could be used to identify? Other CAs? **A**
Streptococcus agalactiae (Group B) is the most common causative agent (although increased awareness is leading to a decrease in incidence of this type of meningitis). CAMP test identifies it. The second most common CA is *Escherichia coli*. Much less common is *Listeria monocytogenes*, which causes abscesses and granulomas in the fetus and a poor prognosis if it crosses the placenta. When acquired at birth, *L. monocytogenes* leads to meningitis but is much less common than *Streptococcux agalactiae* and *E. coli*.

D. Meningitis in babies 6 months to 2 years old who are not vaccinated **Q** CA? Describe current vaccine. **A**
Haemophilus influenzae (Gram–) is still a problem, mostly in unvaccinated children. Vaccine is polyribitol capsule linked to protein.

E. Febrile college-age student is difficult to rouse or comatose and has cutaneous rash. **Q** CA? What population is affected? Describe current vaccine. **A**
The CA is *Neisseria meningitidis*, which is epidemic and can affect any age, although there are larger numbers of cases among 1 year olds and young adults especially in bootcamps, college dorms, and bars. The older vaccine consists of the Y, W-135, C, and A capsular polysaccharides. (YWCA [Young Women's Christian Association] vaccine is used by the military [mostly young men].) The new vaccine is the same polysaccharides conjugated to protein.

F. Meningitis in AIDS and immunocompromised patients **Q** CA in AIDS or immunocompromised (except transplant) patients? CA in transplant patients? **A**
Cryptococcus neoformans (monomorphic yeast) in AIDS or most immunocompromised patients; *Listeria monocytogenes* in transplant patients

G. Meningitis in a severely neutropenic patient Q CA? Describe the organism. A
Aspergillus spp., mainly *Aspergillus fumigatus;* monomorphic filamentous fungus with acute branching

H. Two uncommon bacterial causes of meningitis Q CA if spirochetes with hooked ends? Transmission? CA if acid-fast bacilli are seen? A
Leptospira ("aseptic," transmitted by animal urine in water); *Mycobacterium tuberculosis*

II Brain Abscesses

A. Aerobic, partially acid-fast, Gram+ rods and filaments on microscopy Q CA? A
Nocardia

B. Anaerobic, non-acid-fast filaments and rods seen on microscopy Q CA? A
Actinomyces

III Encephalitis

A. Encephalitis often seen in young adults; high fatality if not treated; cerebrospinal fluid (CSF) near normal; red blood cells (RBCs) in all four tubes of CSF Q CA? Drug of choice? A
Herpes simplex virus (HSV) type 1; acyclovir

B. Types of mosquito-borne encephalitis Q With bird reservoir? Deaths more common in elderly? Deaths more common in young? A
Equine encephalitis (EE); St. Louis encephalitis; California and La Crosse viruses, and West Nile. Elderly most likely to die with St. Louis and West Nile; young, California and La Crosse.

C. Meningoencephalitis with disturbance in smelling; associated with swimming, water skiing, or diving in hot waters Q CA? A
Naegleria

IV Paresthesia

A. Paresthesia with bronze rash or nodular lesions on cool parts of body; culture negative; acid-fast bacteria in lesions Q CA? A
Mycobacterium leprae

V Viruses Latent in Nerves

A. Virus latent in sensory nerve ganglia, with unilateral reactivation involving 1 to 3 dermatomes Q CA? A
Varicella-zoster virus

B. Virus latent in the trigeminal nerve ganglia Q CA? A
HSV-1

C. Virus latent in sacral S-2, S-3 Q CA? A
HSV-2

VI Neurotoxins Produced by Microbes

A. Patient with rigid spasm, trismus (lockjaw), severe spasms on slight noise, opisthotonus, risus sardonicus Q CA? Mechanism of pathogenicity? A
Clostridium tetani; toxin blocks the release of GABA and glycine.

B. Patient (commonly child <1 year) with flaccid paralysis Q CA? Mechanism of pathogenicity? A
Clostridium botulinum; its toxin blocks acetyl choline release.

C. Patient with dysentery and severe headache; has recently immigrated or traveled outside United States Q CA? Describe. CA if acquired in United States? A
Shigella dysenteriae type I (acquired abroad); shiga toxin has neurotoxic activities, in addition to its cytotoxic and enterotoxic activities; shigellae are invasive as well. If in United States, *Shigella sonnei,* which is invasive but produces no toxin.

VII Slow Viral Disease

A. Progressive multifocal encephalopathy Q CA? Viral family? A
JC virus, a polyomavirus belonging to the Papovavirus family (ds DNA with envelope)

B. Subacute sclerosing panencephalitis Q CA? A
Measles virus

VIII Prion Disease

Subacute spongiform encephalopathy Q Human disease names? A
Creutzfeldt-Jakob disease and variant Creutzfeldt-Jakob disease, fatal familial insomnia, and Kuru

Chapter 38

Gastrointestinal and Hepatobiliary Disease

I Vomiting and Diarrhea Caused by Ingestion of Microbial Toxin

A. Vomiting and diarrhea 1 to 6 hours after eating contaminated and poorly refrigerated cream pastries, ham, potato salad; no fever Q Causative agent(s) (CA)? Describe toxin stability. A
Staphylococcus aureus; the enterotoxin (produced in the food) is stable at 60°F (15°C) for at least 10 minutes.

B. Vomiting 1 to 6 hours after eating fried rice Q CA? A
Bacillus cereus, producing emetic toxin

II Noninflammatory Diarrhea (Virus, Noninvasive Bacterium, or Giardia)

A. Watery diarrhea ± vomiting after travel in developing country, but no dramatic dehydration Q CA? A
Enterotoxic *Escherichia coli* (ETEC)

B. Clear diarrhea with mucous flecks ± vomiting after travel in developing country; large fluid loss and rapid dehydration Q CA? What is the virulence factor? A
Vibrio cholerae; cholera toxin is an A-B component toxin; the A component is internalized and ADP-ribosylates G_s alpha, keeping it in a persistent "on" state.

C. Infant or toddler with prolonged watery diarrhea (fall/winter/spring in temperate climate) Q CA? Is it inflammatory, secretory, or malabsorptive? Describe the viral family. A
Rotavirus; secretory diarrhea; naked ds RNA virus

D. Outbreak of nausea, vomiting, and nonbloody, watery diarrhea in five of six family members, 24 to 36 hours after attending a large potluck. Their 9-month-old child is unaffected. Q CA? Why is baby unaffected? What is mode of transmission? A
Norovirus (a.k.a. Norwalk agent); baby did not eat any contaminated food. Transmission from food prepared by sick person or from shellfish from contaminated water.

E. Steatorrheic, foul-smelling diarrhea and intense abdominal cramping in patients who have camped in pristine northern U.S. wilderness camping area and have drunk mountain stream water Q CA? Describe the pathogenic mechanism. Why does milk increase abdominal discomfort? A
Giardia lamblia. Attachment through the ventral sucking disk "coats" the duodenum-jejunum, blunting microvilli; large numbers ultimately cause malabsorption and a transient reduction in lactase levels.

F. Voluminous watery diarrhea with crampy abdominal pain, flatulence, and weight loss; self-limiting in about 2 weeks in immunocompetent people; no effective treatment in immunocompromised patients (may involve biliary tract and may cause death); often acquired through water; acid-fast oocysts seen in intestinal brush biopsy or in diarrhea Q CA? A
Cryptosporidium parvum

G. Abdominal cramps and frankly bloody diarrhea without pus; PMNs not in excess of level in peripheral blood Q CA? Why are antibiotics usually contraindicated, especially in children? A
Escherichia coli 0157:H7 (EHEC) or some other enterohemorrhagic (verotoxic) strain. There is some evidence that antibiotics may increase the risk of hemolytic uremic syndrome, particularly in children.

H. Watery diarrhea lasting 2 to 3 weeks in infant in developing country with fever Q CA? Contrast other most likely infant diarrhea. A
Enteropathogenic *E. coli* (EPEC). Rotavirus is a shorter diarrhea, has a broader age range, and often starts with vomiting.

I. Explosive diarrhea with severe abdominal cramps, vomiting, and fever after ingestion of raw shellfish Q CA? A
Vibrio parahaemolyticus

III Inflammatory Diarrhea and Dysentery

Dysentery or inflammatory diarrheas result when tissue is invaded. USMLE Step 1 cases will usually include basic lab data for the organism. Fever is more prominent than in noninflammatory diarrhea, and mucus will show an excess of PMNs (over levels in peripheral blood).

A. Most common inflammatory diarrhea in United States; the contamination is from improper handling of raw poultry; isolate is oxidase-positive (distinguishing it from any Enterobacteriaceae); it grows at 42°C and may cause a reactive arthritis. Q CA? Describe O_2 intolerance or needs. A
Campylobacter jejuni; it is a microaerophile, so it likes 5% O_2.

B. Inflammatory diarrhea associated with poultry; oxidase-negative, Gram– rod, which is invasive (although the most common species rarely invades blood vessels); the major cause of osteomyelitis in sickle cell disease patients because it has a capsule. Q CA? Most common species? A
Salmonella spp. Most common in diarrhea is *S. enteritidis.*

C. Watery diarrhea, progressing to abrupt onset of febrile disease with abdominal cramps, headache, tenesmus, mucoid stools with or without blood; it may be associated with day care; stools show excess of PMNs, and organism isolated is nonlactose-fermenting, nonmotile Enterobacteriaceae with no animal reservoirs. Q CA? Most common species? Which is most serious and why? A
Shigella. In the United States, *Shigella sonnei* is most common. Most serious is *Shigella dysenteriae* type I, which may progress to hemolytic uremic syndrome because of production of shiga toxin.

D. Abdominal pain, fever, weight loss, usually in association with international travel, ± dysentery (blood and pus in stools) and flask-shaped intestinal lesions; extraintestinal abscesses (especially in liver) are common. Q CA? A
Entamoeba histolytica

E. Antibiotic-associated diarrhea (except amoxicillin-clavulanate, cefixime, and cefoperazone), especially if there are systemic symptoms of fever, colitis, and cramps with leukocytosis and fecal leukocytes Q CA? Pathogenic process? A
Clostridium difficile. Although many antibiotics cause diarrhea, the reduction of intestinal bacteria with the overgrowth of *C. difficile* starts with diarrhea and develops into pseudomembranous colitis (characterized by colitis with cramps, leukocytosis, or fecal leukocytes, and the ultimate development of a yellowish pseudomembrane).

IV Hepatobiliary Disease

A. Review Table 20-2.

B. Infectious causes of cirrhosis including one parasite Q CA? A
Chronic hepatitis B, C, or B/D, but also *Schistosoma mansoni*

C. Bile duct blockage after surgery, fever, or antibiotics Q CA? Why does it happen? A
Ascaris lumbricoides. Blockage occurs because of worm size and the hypermotility under those conditions.

Chapter 39

Cardiovascular Infections, Septicemias, and Blood Cell Changes in Infection

I Cardiac Involvement

A. Native valve endocarditis

1. Acute infective endocarditis Q Causative agent (CA)? Why is it an acute presentation? A

   Staphylococcus aureus. Rapid damage is due to pore-forming alpha toxin.

2. Subacute infective endocarditis is most likely to occur in the elderly or in those with some pre-existing heart disease.

   a. In males after urological manipulations Q CA? A

   Enterococcus faecalis

   b. In people with pre-existing heart damage and very poor oral hygiene or with recent dental work without prophylactic antibiotics Q CA? A

   Viridans streptococci

3. Endocarditis in some homeless persons; no growth on blood agar Q CA? A

   Bartonella henselae or *Bartonella quintana*

4. Endocarditis most common in intravenous drug abusers Q CA and why? What valve is involved most commonly? A

   Staphylococcus aureus, because of abusers' heavy skin flora of *S. aureus;* tricuspid valve. IV drug abusers have many other agents of infective endocarditis; however, because they are less common, they would require additional clues, like a coagulase-negative agent *(S. epidermidis)*.

B. Endocarditis involving a prosthetic valve Q CAs? A

Staphylococcux epidermidis, S. aureus, Enterobacteriaceae, *Pseudomonas, Aspergillus;* questions would also have to give clues about organisms here.

C. Enlarged, flabby heart leading to heart failure, resulting from infection acquired in South or Central America Q CA? Disease? Transmission? What form is found in the heart? A

Trypanosoma cruzi; chronic Chagas disease; transmitted by reduviid bug. The nonflagellated amastigotes are present in the heart muscle.

D. Pericarditis Q CAs? A

Usually viral; most commonly the Coxsackie or other enteroviruses

E. Myocarditis Q CAs? A

Also most commonly the Coxsackie or other enteroviruses, but also seen in Lyme disease

II Septicemias and Shock

Many different infections disseminate via the bloodstream, causing systemic febrile symptoms and septic shock. For example, endotoxin (Gram–), *Staphylococcus aureus* TSST-1, *Streptococcus pyogenes* SPE-A, and peptidoglycan-teichoic acid fragments of Gram+ organisms, such as *Streptococcus pneumoniae,* may all be causative factors. Paper cases will have to give you additional information. Most of the rickettsial diseases, such as Rocky Mountain spotted fever (RMSF), that cause systemic febrile disease will mention capillary and small vessel endothelial damage (and thrombus) and rash and will usually give vector clues (e.g., *Dermacentor* ticks for RMSF). A paper case about brucellosis will include exposure to infected animals.

III Changes in Blood Counts Seen with Infection

A. Anemias

1. Pernicious megaloblastic anemia with history of fish ingestion Q CA? Disease process? A
   Diphyllobothrium latum (fish tapeworm in cool lake regions), attached high in the small intestine, competes for B_{12}, producing anemia.
2. Microcytic hypochromic anemia Q CAs? Disease process? A
   Hookworms *(Necator* or *Ancylostoma)* or *Trichuris*. Blood loss is due to intestinal attachment of a large number of worms.
3. Paroxysmal febrile disease (high fever with chills, rigors, sweats, and headache) that may become cyclic; reduced hematocrit and hemoglobin Q CA? Disease? A
   Plasmodium spp.; malaria

B. Broad generalizations about white cell changes in infection

1. Lymphocytosis with severe hacking cough, usually without fever Q CA? A
   Bordetella pertussis
2. *Eosinophilia* Q When does it occur? A
   During worm migration in tissues or in allergic reactions
3. Mononuclear cell increase Q CA? A
   Intracellular organisms like viruses, *Listeria,* or *Toxoplasma*
4. PMN increases Q CA? A
   Usually extracellular bacteria
5. CD4+ cell decline to <200 per mm^3
   AIDS

Chapter 40

Bone or Joint Infections

I Arthritis

A. Polyarticular arthritis, sometimes migratory, in a 15- to 40-year-old female

1. With petechiae Q Causative agent (CA)? A

Neisseria gonorrhoeae

2. Intermittent with preceding bullet rash and headache or history of tick bite Q Disease? CA? A

Lyme disease; *Borrelia burgdorferi*

B. Nongonococcal infective arthritis, usually monoarticular Q CA? Who is most commonly affected? A

Staphylococcus aureus is most common, especially in individuals with rheumatoid arthritis.

C. Arthritis with artificial joints involved Q CA? A

Staphylococcus epidermidis or *S. aureus*

II Osteomyelitis

A. Osteomyelitis with no comorbidity (including no trauma) Q CA? What is CA when seen in neonates? A

Staphylococcus aureus in all but neonates; in neonates, *S. aureus, S. agalactiae,* and Enterobacteriaceae

B. Osteomyelitis in sickle cell disease patients Q CA? What virulence factor? A

Salmonella (It has a major capsule; *Staphylococcus aureus* is described as having a microcapule.) Sickle cell disease patients have an extremely high rate of osteomyelitis, with more than 80% of cases caused by *Salmonella* spp.

C. Osteomyelitis after a foot wound Q CAs? A

Pseudomonas aeruginosa, Staphylococcus aureus

D. Osteomyelitis involving prosthetic joints Q CAs? A

Staphylococcus aureus, Staphylococcus epidermidis

E. Osteomyelitis in intravenous drug abusers Q CAs? A

Staphylococcus aureus, Pseudomonas aeruginosa

Chapter 41

Genitourinary Tract Infections

I Cystitis

A. In all people (even young sexually active women), most common cause of cystitis Q Causative agent (CA)? A
Escherichia coli

B. In adolescent women who are newly sexually active, second most common cause of cystitis Q CA? What rapid diagnostic lab test would be negative? A
Staphylococcus saprophyticus; nitrite test

II Blood in Urine

A. Symptom of routine bacterial cystitis, but when found in a patient traveling or living in a place like rural Africa Q CA? Disease? Route of acquisition? A
Schistosoma hematobium; schistosomiasis; acquired through water/skin contact

III Reproductive Tract Infections

A. Vesicular genital lesions that ulcerate and recur, with significant nerve pain prior to outbreaks Q CA? Where latent? A
Herpes simplex virus 2; latent in S-2, S-3

B. Genital warts Q CA? Most common serotypes? A
Human papilloma virus; 6 and 11

C. In prepubescent girls, urethritis or vaginitis; in postpubescent females, cervicitis, endometriosis, pelvic inflammatory disease (PID), or perihepatitis; in males, epididymitis; in both sexes, possible Reiter syndrome (urethritis, iridocyclitis, and arthritis, sometimes recurring). Negative culture on any bacteriologic medium. Diagnosis by genetic probe or tissue culture Q CA? Why reinfection? A
Chlamydia trachomatis. Reinfection is common because of multiple serotypes (D–K).

D. In females, asymptomatic condition or urethritis, endocervicitis, or PID; in males, most commonly urethritis. Gram– diplococcus. Q CA? Current test for diagnosis? Culture medium and conditions? A
Neisseria gonorrhoeae; genetic probe tests for diagnosis; culture on Thayer-Martin agar in high CO_2

E. Most commonly seen now in female IV drug abusers or sex workers or in promiscuous male homosexuals. Painless, indurated genital lesions that ulcerate with a fairly clean (not ragged) border. These primary chancres may heal spontaneously, but disease (without treatment) may progress to mucocutaneous lesions on all body surfaces, including palms and soles. Skin lesions are maculopapular; mucous membranes are often grayish, hypertrophic, papular, and more likely to transmit infection than the drier skin lesions. There are also systemic signs, such as fever, lymphadenopathy, malaise, and enlarged spleen. Gram stain of any lesion will be negative (meaning no notable pathogen seen, as opposed to "Gram-negative bacteria seen"). Q CA? Disease? How is diagnosis reached? A
Treponema pallidum; syphilis; dark field or fluorescent microscopy of chancre exudate or mucous membrane secondary lesions; serodiagnosis; not cultured in clinical labs

F. Painful, soft chancre; slow to heal and increases risk of transmission of AIDS Q CA? Describe organism.
Haemophilus ducreyi (You do cry with ducreyi.); Gram– rod

G. In females, asymptomatic or irritating vaginitis with malodorous, thin, yellowish-green discharge (numerous PMNs); discharge pH 5–6 and may give a positive amine test ("whiff" test) when mixed with KOH; vagina congested or with punctate hemorrhages. In males, urethritis. Q CA? A
Trichomonas vaginalis

H. Pruritic, painful vulvovaginitis with normal pH (4–4.5) and pseudohyphae seen on microscopy Q CA? A
Candida albicans

I. Malodorous vaginal discharge; usually worse with intercourse, because ejaculate is basic; usually nonpainful intercourse. Discharge is homogeneous and somewhat adherent and usually gives a positive amine test. Microscopy shows an increased number of coccobacilli. "Clue" cells (epithelial cells covered with bacteria) are present. Q CA? Disease? A
Gardnerella vaginalis at higher levels than normal; bacterial vaginosis

Chapter 42

Cancers Associated with Infections

I Cancers Associated with Viruses

A. Cervical cancer Q Associated agent (AA)? Mechanism of carcinogenesis? How transferred? Associated with what other carcinoma? A
Human papilloma virus (HPV) strains 16 and 18 carry genes that produce early proteins that interfere with normal tumor suppressor gene function. E6 interferes with p53 and E7 with Rb (p110). These are sexually transferred and also associated with penile carcinoma.

B. Burkitt lymphoma Q AAs? A
Epstein Barr virus (EBV) in malarious regions. In a person coinfected with EBV and *Plasmodium,* chromosomal translocation may occur, resulting in lymphoma.

C. Liver carcinoma Q AAs and type of infection? A
Hepatitis B or C chronic infections

D. Human T cell leukemias or lymphomas Q AA? A
Human T cell leukemia virus (HTLV)

II Cancers Associated with Parasites

A. Bladder carcinoma Q AA? A
Schistosoma haematobium chronic infection

B. Plasmodium Q Carcinoma association?
Burkitt lymphoma; as in IB, above

Chapter 43

Skin and Subcutaneous Infections and Rashes

I Skin and Subcutaneous Infections

A. Surgical wounds or carbuncles (boils) or furuncles (multiple ulcerating boils) Q Causative agent (CA)? A
Staphylococcus aureus

B. Swollen jaw following dental trauma or work; may ulcerate to the surface with the presence of "sulfur" granules in the exudate Q CA? Description of causative agent? What are the granules? A
CA: *Actinomyces israelii,* a bacterium! It is a Gram+ anaerobe that is not acid-fast. Granules are microcolonies.

C. Dermatitis along tight areas of swimsuit after swimming Q Name of ailment? CA? A
Swimmer's itch from infection with "nonhuman" schistosomes

D. Itchy skin lesions spreading out from the periphery; margins usually erythematous; appear anywhere on body; KOH shows hyphae and arthroconidia. Q Type of infection? Specific causative agents and tissues each agent infects? A
Dermatophyte infection (tinea):

- *Microsporum* invades hair and skin.
- *Trichophyton* invades hair, nails, and skin.
- *Epidermophyton* invades skin and nails.

E. Impetigo caused by a Gram+ coccus positive for both catalase and coagulase Q CA? Typical clinical appearance? A
Staphylococcus aureus; may have bullae (or be described as bullous)

F. Swollen subcutaneous lesion (or chain of lesions) resulting from trauma involving plant materials, such as plum tree, rose, or other thorny plant, or floral wire; occupational hazard of greenhouse and plant nursery workers and gardeners Q Disease? CA? Description of CA in environment versus in humans? A
Disease: Sporotrichosis. *Sporothrix schenckii,* the CA, is a dimorphic fungus that is filamentous in the environment; in human tissues, it forms sparse, cigar-shaped and oval yeasts.

G. A primary concern with deep puncture wounds or any complex dirty wound Q CA? Most common infectious form? Mechanism of pathogenicity? A
Clostridium tetani; spores and vegetative cells; neurotoxin production causes rigid paralysis.

H. Cellulitis caused by nail through bottom of tennis shoe Q CA? Where is organism coming from? A
Pseudomonas aeruginosa is the most common CA. The organism comes from the inside of the tennis shoe (rubber sole "cleans up" outside dirt on nail).

I. Wound contaminated with fecal material or soil, leading to the production of gas trapped in tissues and gangrene Q CA? Major toxin? A
Clostridium perfringens; alpha toxin, a lecithinase

J. Necrotizing fasciitis Q CA? A
Streptococcus pyogenes

K. Superficial peeling of large areas of skin Q CA? Disease? A
Staphylococcus aureus; scalded skin syndrome due to exfoliatins

L. Diaper rash with sharply demarcated red to normal skin, with a few punctate red dots on the normal skin Q CA? What would be seen on microscopy? A
Candida spp.; pseudohyphae, hyphae, and yeasts on microscopy

M. Impetigo caused by a catalase-negative organism, usually described as a honey-crusted lesion Q CA? A
Streptococcus pyogenes

N. Burned tissue with blue-green pus and odd, sweet, grapelike odor Q CA? A
Pseudomonas aeruginosa

O. Dermatitis with fairly intense itching that develops after walking barefoot on a tropical beach Q Most likely CA? Disease? Other CAs? A
CA: Dog ascarids that infect humans. Disease: cutaneous larva migrans. Other causes include repeated hookworm *(Necator* or *Ancylostoma)* or *Strongyloides* infections from walking barefoot on soil.

P. Pink umbilicate warts with central debris Q CA? Pathology clue? Viral family? A
CA: Molluscum contagiosum virus. Clue: molluscum bodies in cell cytoplasm. Viral family: Poxviridae.

Q. Patient working with animal, animal skins, or wool develops a red tumorlike lesion with central necrosis and a raised red margin Q CA? Disease? Unique virulence factors? A
CA: *Bacillus anthracis*. Disease: anthrax. Virulence factors: polypeptide capsule and two toxins (LF, EF) that share a B component (protective antigen, or PA).

II Rashes

In general, vesicular rashes are more often viral than bacterial; *Staphylococcus* may cause vesicles or bullae.

A. Vesiculate rash starts in hair and behind ears and spreads downward, with major concentration on trunk and sparse distribution on limbs. The rash is asynchronous (the initially vesicular lesion quickly ulcerates and crusts over, and there are progressive new crops of lesions). Fever ranges up to 103°F (39.4°C) for a few days after the rash starts. Q CA? Disease A
Varicella zoster virus; chickenpox

B. Seasonal infection (from deep winter to early spring), beginning with cough, coryza, and conjunctivitis, with fever ranging from 101°F to 103°F (38°C to 39.4°C) prior to the onset of rash. (White oral buccal lesions with a red base appear on days 3 to 6 but often are not noted.) Blotchy red rash starts on face and spreads downward, becoming confluent on face and upper trunk, while remaining discrete on lower extremities Q Disease (common and formal names)? CA? A
Measles (rubeola); measles virus

C. Painful outbreak of vesiculate lesions along 1 to 3 dermatomes Q CA? Disease? A
Varicella zoster virus; shingles

D. Symptoms include sore throat, often with yellowish tonsillar exudate; strawberry tongue; fever; and sparse, fine, "sandpaper" rash that blanches on pressure and usually starts on cheeks, sparing circumoral area, with increased density at neck, axillae, and groin Q Disease? CA? Virulence factor? A
Disease: scarlet fever. CA: *Streptococcus pyogenes*. Virulence factor: SPE-A, B, or C is the erythrogenic toxin.

E. Teen to young adult with exudative sore throat, fever, lymphadenopathy, and fatigue; when given ampicillin, develops major rash with no change in throat Q CA? Describe CA. A
Epstein Barr virus; enveloped DNA virus (icosahedral)

F. Febrile child, not very sick, with red cheeks and thin lacy rash on body; condition usually seen in late spring Q CA? Disease? A
Parvovirus B19; fifth disease ("slapped cheek fever")

G. Child 5 months to 3 years with high fever for 3 days; maculopapular facial rash appears after fever abates Q CA? Disease? A
Human herpesvirus 6; exanthem subitum (roseola infantum)

H. Acute onset of illness: healthy person suddenly becomes very ill, with mental confusion and petechiae very quickly developing into purpura and shock Q CA? Disease? A
Neisseria meningitidis, causing meningococcemia and meningitis

Part VII

Comparative Microbiology

This section is a "review of the review," providing horizontal comparisons of microbes, features, virulence factors, and so on. As in Part VI, you can test yourself by using a cover sheet and reading down line by line until you hit a list heading. Then, on a separate sheet of paper, use abbreviations to quickly write your own list. Compare your list with the one given, then add to your list anything you missed. This, along with the flowcharts in Chapter 7, makes a good last-minute review. If you just want to read through these lists, be sure to make a separate list of the items you don't know for later study.

Chapter **44**

High-Yield Microbial Clues and Comparisons

I Virulence Factors and Features Important in Disease

A. SURFACE ADHERENCE AND COLONIZATION

1. Gram-positive teichoic acids
2. M proteins of *Streptococcus pyogenes* fimbriae
3. Gram-negative pili
4. Immunoglobulin A (IgA) proteases: *Streptococcus pneumoniae, Neisseria* spp., *Haemophilus influenzae*
5. Biofilms
   a. *Staphylococcus epidermidis* biofilms adhere to artificial body parts and catheters.
   b. *Streptococcus mutans* biofilms cause dental plaque.

B. ANTIPHAGOCYTIC STRUCTURES

1. **All capsules.** Capsules are the most important antiphagocytic structures. The following is a mnemonic for high-yield **polysaccharide capsules:** Ⓜ S^{2}ome K^{2}illers Have Pretty Nice Capsules:
   a. *Streptococcus pneumoniae* and *Streptococcus agalactiae*
   b. *Klebsiella pneumoniae* and K1 strains of *Escherichia coli*
   c. *Haemophilus influenzae*
   d. *Pseudomonas aeruginosa*
   e. *Neisseria meningitidis*
   f. *Cryptococcus neoformans* (yeast)
   Remember that *Bacillus anthraces, Francisella tularensis,* and *Yersinia pestis* all have polypeptide capsules that are also antiphagocytic.
2. **Pili** of *Neisseria gonorrhoeae*
3. **M protein** of *Streptococcus pyogenes*
4. **A protein** of *Staphylococcus aureus*

C. TOXINS

1. **Toxins with ADP ribosyl transferase activity** are important in disease. Table 44-1 lists high-yield examples.
2. **Heat-stable bacterial toxins**
   a. Boiling point of water **(100°C; 212°F): endotoxin**
   b. **60°C (140°F)** for 10 minutes

TABLE 44-1 TOXINS WITH ADP RIBOSYL TRANSFERASE AND RESULTING CONDITIONS

Toxin	ADP Ribosylates	Result
Cholera toxin	G_s alpha protein of small intestinal cells	Stimulates G_s, increase in cAMP Massive fluid loss (diarrhea)
E. coli heat-labile toxin (LT)	G_s protein of small intestinal cells	Stimulates G_s, increase in cAMP Fluid loss (diarrhea)
Pertussis toxin	G_i (inhibits the negative regulator of adenylate cyclase)	Inhibits G_i Increase in cAMP, lymphocytosis, increased insulin secretion
Diphtheria toxin	EF-2	Shuts down protein synthesis
Pseudomonas aeruginosa exotoxin A	EF-2	Shuts down protein synthesis

KEY: cAMP, cyclic adenosine monophosphate; EF-2, elongation factor 2.

i. ***Staphylococcus aureus* enterotoxin**
ii. **Stable toxin (ST) of *Escherichia coli***
iii. *Yersinia enterocolitica* toxin

c. Toxins denatured by moderate heat: the remaining bacterial toxins

3. **Toxins that cause membrane damage:** *Clostridium perfringens* alpha toxin (a lecithinase), *Staphylococcus aureus* alpha toxin (a pore-forming toxin), *Listeria monocytogenes* (Listeriolysin O facilitates endosome escape.)
4. **Other toxins:** For review, see toxins in Table 4-1.

D. OTHER VIRULENCE FACTORS

1. **Coagulase:** *Staphylococcus aureus* and *Yersinia pestis*
2. **Urease:** *Cryptococcus neoformans, Nocardia,* ***Helicobacter pylori*** (allows survival in lumen of stomach; the produced ammonia adds to inflammation), ***Proteus,*** and *Ureaplasma* (The last two, especially *Proteus,* increase urine pH in urinary tract infections, causing kidney stones.)
3. **Hyaluronidase:** Group A streptococci and *Staphylococcus aureus*

II Identifiers of Microorganisms

A. STAIN REACTIONS

1. **Not reliably seen on Gram stain**
   a. *Mycoplasmas, Ureaplasma, Chlamydia;* no peptidoglycan
   b. *Chlamydia, Rickettsia;* too small
   c. Spirochetes (*Treponema, Borrelia, Leptospira*); too thin
   d. *Legionella;* Gram-negative stain seen only if counterstain time is increased.
2. **Acid fast**
   a. *Mycobacterium*
   b. *Nocardia* (partially)
   c. *Legionella micdadei*
   d. *Cryptosporidium oocysts*
   e. *Isospora oocysts*

3. **Silver staining**
   a. Fungi
   b. *Legionella*
4. **Periodic acid-Schiff:** Fungi stain red.
5. **Calcofluor white:** Fungi fluoresce blue-white on black.
6. **India Ink wet mount:** *Cryptococcus* shows colorless cells with halos (capsular material) on black; test misses 50%.

B. SPECIAL CULTURE LIMITATIONS

1. **Intracellular pathogens:** require tissue culture; cannot be grown on inert media or in vitro.
   a. **Obligate intracellular:** all viruses; *Chlamydia;* all Rickettsias, except *Bartonella (Rochalimaea); Mycobacterium leprae; Plasmodia; Toxoplasma*
   b. **Facultative intracellular:** *Listeria,* all Mycobacteria, *Histoplasma, Brucella*
2. **Nonintracellular pathogens** that **cannot** yet be **routinely cultured:** *T. pallidum, Pneumocystis*

C. ENDOSPORE FORMERS (Gram-positive, dipicolinic acid in core): *Bacillus, Clostridium*

D. ANAEROBES. Don't worry about obligate versus aerotolerant for the USMLE Step 1 exam. The ABCs of anaerobiosis are *Actinomyces, Bacteroides, Clostridium.*

E. MICROAEROPHILIC ORGANISMS: *Campylobacter* grows at 42°C (107.6°F); *Helicobacter,* at 37°C (98.6°F).

F. OBLIGATE AEROBES: *Mycobacterium tuberculosis, Pseudomonas*

G. MEDIA CLUES AND GROWTH FACTORS are listed in Table 44-2. Quiz yourself by covering one side of the table at a time.

TABLE 44-2 MEDIA CLUES OR GROWTH FACTORS AND ORGANISMS FOR WHICH THEY ARE USED

Media Clue or Growth Factor	Used for Which Organism(s)?
Charcoal yeast extract agar	*Legionella*
Cholesterol	Mycoplasmas and Ureaplasmas
Salt tolerant	*Staphylococcus aureus, Enterococcs faecalis, Streptococcus bovis, Vibrio parahaemolyticus, Vibrio vulnificus*
X and V factors (hematin and NADH)	*Haemophilus influenzae*
Chocolate agar	*Haemophilus* and *Neisseria*
Thayer-Martin (chocolate with antibiotics)	*Neisseria* (from nonsterile body sites)
Regan-Lowe	*Bordetella pertussis*
Lowenstein-Jensen (or media with palmitic acid)	Mycobacteria (automated broth systems)

TABLE 44-3 HIGH-YIELD GUIDE TO DISEASES AND THEIR CAUSATIVE AGENTS, RESERVOIRS, AND MODES OF TRANSMISSION

Disease	Causative Agent	Reservoir	Mode of Transmission
All STDs	*Treponema pallidum, Neisseria gonorrhoeae, Chlamydia* HPV, HSV-2, etc.	Humans only	Sexual, birth, or other direct contact; *T. pallidum* crosses the placenta
Typhoid	*Salmonella typhi*	Humans only	Fecal-oral
Leprosy	*Mycobacterium leprae*	Humans (also armadillos, mangabey monkeys)	Respiratory droplets and direct contact
Cryptococcal meningitis (AIDS, Hodgkin)	*Cryptococcus neoformans*	Weather, pigeon droppings	Inhalation (often reactivation)
African sleeping sickness	*Trypanosoma brucei*	Humans	Tsetse flies
Histoplasmosis	*Histoplasma capsulatum*	Soil; bird or bat feces in U.S. great river valleys	Inhalation
Infectious hepatitis	Hepatitis A virus	Humans, water, oysters	Fecal-oral or food-borne
Bloody diarrhea with excess fecal leukocytes	*Campylobacter*	Domestic animals	Undercooked chicken
Bloody diarrhea *without* excess fecal leukocytes	*Escherichia coli* O157:H7	Cattle	Undercooked hamburger
Salmonellosis	*Salmonella*	Chicken, eggs	Undercooked chicken
Chagas disease	*Trypanosoma cruzi*	Domestic animals and pets alternating with reduviid bugs	Reduviid bugs ("cone" or "kissing" bugs)
Valley fever	*Coccidioides immitis*	Desert sand of southwestern United States	Dust (with arthroconidia) inhalation
Malaria	*Plasmodium* spp.	Mammals alternating with *Anopheles* mosquitoes	*Anopheles* mosquito bite
Leishmaniasis (all types)	*Leishmania* spp.	Sand flies	Sand fly bite
Lyme disease	*Borrelia burgdorferi*	Deer, deer mice, *Ixodes* ticks	*Ixodes* tick bite (nymph or adult)
Leptospirosis	*Leptospira interrogans*	Rats, cattle	Animal urine in recreational waters or sewers
Toxoplasmosis	*Toxoplasma gondii*	Cats and all other animals, including humans	Raw meat, cat feces; crosses placenta, reactivates in immunocompromised patients
Rocky Mountain spotted fever	*Rickettsia rickettsii*	Dogs and ticks	*Dermacentor* ticks
Legionnaire disease	*Legionella* spp.	Water and amebas	Air conditioning
Viral diarrheas	Rotavirus (infants), Norovirus (older persons)	Humans; may also be shellfish for Norovirus	Fecal-oral
Gray, greasy diarrhea	*Giardia lamblia*	Humans and beavers	Water, fecal-oral
Schistosomiasis	*Schistosoma* spp.	Snails are alwaysone host	Water contact
Plague	*Yersinia pestis*	Small wild animals in southwestern U.S. deserts	Flea bite, or respiratory droplets
Tuberculosis	*Mycobacterium tuberculosis*	Humans	Respiratory droplets or droplet nuclei

KEY: HPV, human papillomavirus; HSV-2, herpes simplex virus type 2.

H. SPECIFIC INHIBITORS

1. Bacitracin ("A" disk used by labs) inhibits *Streptococcus pyogenes* (Group A).
2. Optochin ("P" disk used by labs) inhibits (and bile lyses) *Pneumococcus*.

I. The following are the most likely to be pictured or physically described on the USMLE Step 1 exam:

1. *Mycobacterium:* thin red rods on acid-fast stain (aerobic)
2. A Gram-positive organism: intense blue to purple (tissues pale red)
3. A Gram-negative organism: pink to clear red
4. Adenovirus: icosahedral, projecting fibers. (See Table 21-1.)
5. *Giardia:* flagellated pyriform protozoan with a sucking disk
6. *Entamoeba histolytica:* ameba with "wagon wheel" nucleus and ingested red blood cells (RBCs)
7. Dimorphic fungi: in body, yeast or yeastlike; at lower temperatures (cold), mold
   a. *Blastomyces dermatitidis:* broad-based budding yeast or hyphae; possible association with wood
   b. *Coccidioides immitis:* hyphae and arthroconidia in sand from southwestern United States; in lungs, arthroconidia develop into spherules.
   c. *Histoplasma capsulatum:* hyphae with tuberculate macroconidia and microconidia in soil enriched with bird or bat feces; in tissue, facultative intracellular fungus seen as tiny oval budding yeasts
   d. *Sporothrix schenckii:* filamentous hyphae (rosettes or sleeves of conidia) on plant material; in tissue, develops into oval to cigar-shaped yeasts

III High-Yield Transmission Data

Table 44-3 lists important diseases and their causative agents, reservoirs, and modes of transmission. Although fecal organisms may be transmitted by oral sex or from exposed feces by flies, the usual route is from fingers to food or from fingers to fomites as in day-care situations (a major route for fecal-oral transmission).

Part **VIII**

High-Yield Case Setups

This section of high-yield case setups and related questions is designed to be used as a self-test so you can determine whether you need more review. Each of the 22 items consists of a high-yield case scenario, followed by questions about the case; answers to the questions then follow. Only critical clues are given in each case. Item 4 includes case modifications that require you to select a different causative agent.

Directions: Use a cover sheet so you can read down the page one line at a time. A question icon (Q) separates the case information in each item from the questions about the case. When you see the answer icon (A), stop and answer the questions, then move your cover sheet down to check your answers before moving on to the next item.

A note about causative agents (CAs): The clinical vignette questions on the USMLE Step 1 exam often do not ask for the causative agent, but you must know this information in order to answer basic science questions about the disease or the causative agent. For this reason, every case in this appendix asks you to identify the causative agent.

A note about the use of *and* versus *or:* This self-test, like the USMLE Step 1 exam, uses the words *and* and *or* in the Boolean sense—that is, the word *or* between clues means that any one of these clues will affect your answer, whereas the word *and* means all clues apply.

Chapter **45**

High-Yield Case Setups

1. Lobar pneumonia in an alcoholic (or elderly) patient Q CA? Chemistry and function of major virulence factor? Sputum? Lab ID of CA? A
CA: *Streptococcus pneumoniae* (by far the most common CA in alcoholics and the elderly.) Major virulence factor: The polysaccharide capsule reduces complement activation and phagocytic engulfment. Sputum: Blood tinged or rusty but not foul-smelling Lab ID: CA is alpha-hemolytic, inhibited by optochin (P disk), and lysed by bile (bile activates *S. pneumoniae* autolysins).

2. Lobar pneumonia with deep red (currant jelly) sputum or Gram-negative CA. Q CA? Which patients are most likely to have it? Associated symptom that makes treatment difficult? A
CA: *Klebsiella pneumoniae* Likely patients: Alcoholics and patients with chronic obstructive pulmonary disease (COPD) Associated symptom: This type of pneumonia is very serious because of the high frequency of abscesses, but it is not as common as pneumonia caused by *S. pneumoniae*.

3. Pneumonia in an alcoholic, with or without vomiting and passing out or with foul-smelling sputum; with or without pulmonary abscesses, sometimes with visible fluid level. Q CA? Where are they from? A
Mixed anaerobes from aspiration of oral/pharyngeal normal flora and vomitus

4. Meningitis in neonate born after long delivery, with early rupture of amniotic sac Q Most common CA? What factor increases risk? A
Most common CA: *Streptococcus agalactiae* (Group B *Streptococcus*) Major risk: Prolonged membrane rupture
   a. Modification 1: Meningitis occurs in neonate and CA is Gram-negative. Q CA? A
   Escherichia coli
   b. Modification 2: Meningitis occurs in neonate and CA is Gram-positive rod. Q CA? What other patient population is sensitive to this organism? A
   Listeria monocytogenes; more likely to cause meningitis in transplant patients

5. Foot swelling and redness as a result of stepping on nail that penetrates tennis shoe Q CA? What other diseases does this CA cause in other patient populations? A
Pseudomonas aeruginosa. Also causes pneumonia in cystic fibrosis or severely neutropenic patients; cellulitis in burn patients; folliculitis in hot tub users; malignant otitis media in diabetics; eye ulcers in coma patients or patients using extended wear contacts

6. Knee joint pain and swelling and skin petechiae with or without low-grade fever in a sexually active young woman Q CA? Disease? What is immunity or reinfection potential following treatment? How is diagnosis made? If cultured, what medium? A

CA: *Neisseria gonorrhoeae* Disease: Arthritis from disseminated and untreated gonorrhea Reinfection: Common antigenic variation of pili and outer membrane proteins results in high risk of reinfection of a person with multiple sexual partners who does not use barrier protection. Diagnosis and culture medium: Gonorrhea diagnosis is made with use of genetic probes. *Neisseria gonorrhoeae* is cultured on Thayer-Martin, because it will not grow well on blood agar.

7. Patient (frequently child) has acute pharyngitis with yellowish exudate, regional lymphadenopathy, often high fever; beta-hemolytic organism inhibited by bacitracin Ⓠ CA? Potential suppurative sequelae if untreated? Potential nonsuppurative sequelae? Ⓐ
CA: *Streptococcus pyogenes* Suppurative sequelae: Otitis media, sinusitis, abscesses Nonsuppurative sequelae: Rheumatic fever (polyarthritis of large joints, carditis, erythema nodosum and marginatum, chorea) or acute glomerulonephritis

8. Vietnamese person escaped on a boat and spent 4 years in a Pacific refugee camp before coming to United States; now suffers from back pain, weight loss, and night sweats; x-ray film shows vertebral destruction Ⓠ CA and description? Growth medium? Ⓐ
Mycobacterium tuberculosis is an acid-fast bacterium requiring a high-lipid medium to grow (Lowenstein-Jensen medium), although it is now also grown in Middlebrook broth with radioactive palmitic acid in special machines.

9. Febrile young person with rapid onset of stiff neck, mental confusion, and skin petechiae Ⓠ CA? What surface component enables CA to *reach* the blood-brain barrier (BBB)? What enables it to *cross* the BBB? Which serotype is least immunogenic? Is it contagious? Ⓐ
CA: *Neisseria meningitidis* most likely. Surface components: Polysaccharide capsule protects the bacterium and enables it to reach the BBB. Endotoxin in the outer membrane causes inflammation that facilitates invasion of the central nervous system (CNS). Least immunogenic serotype: B capsule—composed of sialic acid, so it is not immunogenic Contagiousness: It is an epidemic disease, although many more people are colonized than develop disease.

10. Severely neutropenic patient with respiratory distress; bronchioalveolar lavage fluids show septate hyphal elements (filaments) with generally acute, dichotomous branching Ⓠ CA? Ⓐ
Aspergillus fumigatus; a monomorphic filamentous fungus

11. Immigrant from lower socioeconomic class of South or Central America presents with fatigue and an enlarged heart; biopsy shows amastigotes Ⓠ CA? Disease? How acquired? How does it affect adults? What country is most affected? Why is it a concern in the United States? Ⓐ
CA: *Trypanosoma cruzi* Disease: Trypanosomiasis (Chagas disease) Acquired: In poorer housing from bites of reduviid bugs (cone or kissing bugs). In adults: Infections are often chronic and lead to heart failure. Countries: It is a major cause of death in Brazil and is a transfusion concern in the United States.

12. Patient presents with bone pain and tender, inflamed tissues over the painful area; no history of trauma; no underlying disease; patient not a neonate Ⓠ CA? Type of organism? Reactions to coagulase and catalase? Ⓐ
CA: *Staphylococcus aureus*, a Gram-positive coccus. Reactions: Coagulase-positive, catalase-postive

13. Osteomyelitis in sickle cell disease patient Q CA? Why are these patients susceptible? A *Salmonella enteritidis;* Sickle cell disease patients (who are asplenic from repeated infarcts and also may have a defect in complement) do not control encapsulated organisms well, so they have a very high rate of osteomyelitis. (*Staphylococcus aureus* is rarely encapsulated.)

14. Profuse watery diarrhea leading to rapid and severe dehydration Q CA? Pathogenesis? Management? A
CA: *Vibrio cholerae* Pathogenesis: Toxin is ADP ribosyl transferase of G-binding protein, causing increase in cAMP. Management: High risk of hypovolemic shock unless treated with electrolytes and fluid replacement. Antibiotics reduce spread.

15. Adoptee from former Soviet Union has pharyngitis with pseudomembrane formation Q CA? Other organs likely to be involved? Pathogenesis? Definitive test? Culture medium or other distinctive clues likely to be mentioned? Prevention (type of vaccine)? A
CA: *Corynebacterium diphtheriae* Other organs: Heart (myocarditis/heart failure) and CNS (recurrent laryngeal palsy) Pathogenesis: Circulating exotoxin inhibits protein synthesis through ADP ribosylation of elongation factor 2 (EF-2). Definitive test: Elek test (agar immunodiffusion to identify toxin production) Culture medium: Loeffler coagulated serum medium and tellurite-containing medium are distinctive. Vaccine: Toxoid

16. Woman with cervical intraepithelial neoplasia Q CA and serotypes? Pathogenic mechanism? A
CA and serotypes: Human papilloma virus 16 or 18 Pathogenic mechanism: Early proteins interfere with tumor suppressor gene activity—E6 interferes with p53; E7, with p110 (Rb).

17. Wilderness camper who drank mountain stream water presents with abdominal pain and gray steatorrheic stools Q CA? How does it cause diarrhea? A
Giardia lamblia fastens to duodenal and jejunal lining by ventral sucking disk, blocking absorption.

18. Hospitalized patient with enterococcal septicemia, being treated with clindamycin, develops diarrhea on day 5 Q CA? Appropriate tests? Pathogenesis? A
CA: *Clostridium difficile* Tests: Do not order ova and parasites; test for presence of *C. difficile* toxins A and B. Pathogenesis: Through exotoxin production; toxin A is an enterotoxin and granulocyte attractant, causing mucosal damage and water and electrolyte loss. Toxin B is cytotoxin.

19. College student camping through tropical developing country develops watery (nonbloody) diarrhea Q Most likely CA if self-resolving after a couple of days of saltines and 7-Up? Pathogenesis? A
CA: Enterotoxic *E. coli.* Pathogenesis: Production of two possible exotoxins: LT ADP ribosylates G_s, activating an adenylate cyclase; ST activates a guanylate cyclase.

20. College student camping through tropical third-world country develops abdominal pain and bloody diarrhea. Patient is febrile and liver is enlarged. Diarrhea contains inflammatory cells, cysts, and motile ameboid trophozoites with ingested red blood cells. Nucleus has a sharp central karyosome. Q CA? Description of the pathology? A
CA: *Entamoeba histolytica* Pathology: The organism is highly invasive and produces flask-shaped ulcers responsible for the characteristic extraintestinal spread.

21. Medical student complains of severe fatigue of 2 weeks' duration and sore throat; physical findings include fever, palpable spleen, postauricular cervical lymphadenopathy, and exudative pharyngitis. Ⓠ CA? Which antibodies give positive screening test, and with what do they cross-react? Ⓐ
Epstein-Barr virus; heterophile antibodies cross-react with animal red blood cells, not with the virus.

22. Febrile neonate born after extended rupture of membranes; mother also febrile. Infant has Apgar scores of 2 and 7 at 1 and 5 minutes, respectively. Ⓠ CA? Ⓐ
Streptococcus agalactiae

Unnumbered Figure Acknowledgements

Chapter 1

(left) Electron micrograph of fibrillar prion protein; modified from Figure 34.2. Strohl, W.A., et al. *Lippincott's Illustrated Reviews: Microbiology*. R.A. Harvey and P.A. Champe, eds. Philadelphia: Lippincott Williams & Wilkins, 2001, p. 397, with permission from Elsevier.

(right) Mating pair of schistosomes; modified from Figure 25.4. Strohl, W.A., et al. *Lippincott's Illustrated Reviews: Microbiology*. R.A. Harvey and P.A. Champe, eds. Philadelphia: Lippincott Williams & Wilkins, 2001, p. 291.

Chapter 2

(left) Gram-positive bacterial cell envelope; modified from Figure 2-1. Hawley, L., et al. *High-Yield Microbiology and Infectious Diseases*. Philadelphia: Lippincott Williams & Wilkins, 2000, p. 10.

(right) Gram-negative bacterial cell envelope; modified from Figure 2-2. Hawley, L., et al. *High-Yield Microbiology and Infectious Diseases*. Philadelphia: Lippincott Williams & Wilkins, 2000, p. 11.

Chapter 3

Growth curve; modified from Figure 3-1 top. Hawley, L., et al. *High-Yield Microbiology and Infectious Diseases*. Philadelphia: Lippincott Williams & Wilkins, 2000.

Chapter 4

(left) A cytolytic toxin; modified from Figure 11.3. Strohl, W.A., et al. *Lippincott's Illustrated Reviews: Microbiology*. R.A. Harvey and P.A. Champe, eds. Philadelphia: Lippincott Williams & Wilkins, 2001, p. 139.

(right) A superantigen toxin; modified from Figure 11.3. Strohl, W.A., et al. *Lippincott's Illustrated Reviews: Microbiology*. R.A. Harvey and P.A. Champe, eds. Philadelphia: Lippincott Williams & Wilkins, 2001, p. 139.

Chapter 5

(left) Bacteriophage Lambda; modified from Figure 5-7. Hawley, L., et al. *High-Yield Microbiology and Infectious Diseases*. Philadelphia: Lippincott Williams & Wilkins, 2000, p. 33.

(right) Generalized transducing Bacteriophage Lambda with *E. coli* DNA; modified from Figure 5-7. Hawley, L., et al. *High-Yield Microbiology and Infectious Diseases*. Philadelphia: Lippincott Williams & Wilkins, 2000, p. 33.

Chapter 6

(left) Antibiotic resistance: alteration of target; modified from Figure 10.17. Strohl, W.A., et al. *Lippincott's Illustrated Reviews: Microbiology*. R.A. Harvey and P.A. Champe, eds. Philadelphia: Lippincott Williams & Wilkins, 2001, p. 133.

(middle) Antibiotic resistance: decreased permeability; modified from Figure 10.17. Strohl, W.A., et al. *Lippincott's Illustrated Reviews: Microbiology*. R.A. Harvey and P.A. Champe, eds. Philadelphia: Lippincott Williams & Wilkins, 2001, p. 133.

(right) Antibiotic resistance: inactivation of drug; modified from Figure 10.17. Strohl, W.A., et al. *Lippincott's Illustrated Reviews: Microbiology*. R.A. Harvey and P.A. Champe, eds. Philadelphia: Lippincott Williams & Wilkins, 2001, p. 133.

Chapter 7

(left) *Streptococcus pneumoniae;* modified from Figure 12.10. Strohl, W.A., et al. *Lippincott's Illustrated Reviews: Microbiology*. R.A. Harvey and P.A. Champe, eds. Philadelphia: Lippincott Williams & Wilkins, 2001, p. 151.

(middle) CSF with *Neisseria meningitidis;* modified from Figure 14.9. Strohl, W.A., et al. *Lippincott's Illustrated Reviews: Microbiology*. R.A. Harvey and P.A. Champe, eds. Philadelphia: Lippincott Williams & Wilkins, 2001, p. 169, with permission from Images in clinical medicine. *N Engl J Med.* 1997;336:707.

(right) Dark-field microscopy of *Treponema pallidum;* modified from Figure 18.3. Strohl, W.A., et al. *Lippincott's Illustrated Reviews: Microbiology*. R.A. Harvey and P.A. Champe, eds. Philadelphia: Lippincott Williams & Wilkins, 2001, p. 222.

Chapter 8

(left) *Staphylococcus aureus;* modified from Figure 11.6. Strohl, W.A., et al. *Lippincott's Illustrated Reviews: Microbiology*. R.A. Harvey and P.A. Champe, eds. Philadelphia: Lippincott Williams & Wilkins, 2001, p. 141.

(middle) *Streptococcus pneumoniae;* modified from Figure 12.10. Strohl, W.A., et al. *Lippincott's Illustrated Reviews: Microbiology*. R.A. Harvey and P.A. Champe, eds. Philadelphia: Lippincott Williams & Wilkins, 2001, p. 151.

(right) *Enterococcus faecalis;* modified from Figure 12.5. Strohl, W.A., et al. *Lippincott's Illustrated Reviews: Microbiology*. R.A. Harvey and P.A. Champe, eds. Philadelphia: Lippincott Williams & Wilkins, 2001, p. 154.

Chapter 9

(top left) *Clostridium difficile;* modified from Figure 17.2. Strohl, W.A., et al. *Lippincott's Illustrated Reviews: Microbiology*. R.A. Harvey and P.A. Champe, eds. Philadelphia: Lippincott Williams & Wilkins, 2001, p. 210, with permission from Bottone, E.J., Girolami, R., and Stamm, J.M., Schneierson's *Atlas of Diagnostic Microbiology,* 9th ed. North Chicago: Abbott Laboratories, 1984.

(top middle) *Listeria monocytogenes;* modified from Figure 13-10. Strohl, W.A., et al. *Lippincott's Illustrated Reviews: Microbiology*. R.A. Harvey and P.A. Champe, eds. Philadelphia: Lippincott Williams & Wilkins, 2001, p. 162. Copyright A.M. Siegelman/Visuals Unlimited.

(top right) *Corynebacterium diphtheriae;* modified from Figure 13-2. Strohl, W.A., et al. *Lippincott's Illustrated Reviews: Microbiology*. R.A. Harvey and P.A. Champe, eds. Philadelphia: Lippincott Williams & Wilkins, 2001, p. 158, with permission from Elsevier.

(bottom left) *Actinomyces* spp.; modified from Figure 21.21. Strohl, W.A., et al., *Lippincott's Illustrated Reviews: Microbiology*. R.A. Harvey and P.A. Champe, eds. Philadelphia: Lippincott Williams & Wilkins, 2001, p. 257, with permission from Alexander, S.K. and Strete, D. *Microbiology: A Photographic Atlas for the Laboratory*. Copyright by Benjamin Cummings, 2001. Reprinted with permission from Pearson Education.

(bottom middle) *Nocardia* spp.; modified from Figure 21.23. Strohl, W.A., et al. *Lippincott's Illustrated Reviews: Microbiology*. R.A. Harvey and P.A. Champe, eds. Philadelphia: Lippincott Williams & Wilkins, 2001, p. 257.

(bottom right) *Mycobacterium tuberculosis;* modified from Figure 21.2. Strohl, W.A., et al. *Lippincott's Illustrated Reviews: Microbiology*. R.A. Harvey and P.A. Champe, eds. Philadelphia: Lippincott Williams & Wilkins, 2001, p. 246.

Chapter 10

Mycoplasma spp.; modified from Figure 19.3. Strohl, W.A., et al. *Lippincott's Illustrated Reviews: Microbiology*. R.A. Harvey and P.A. Champe, eds. Philadelphia: Lippincott Williams & Wilkins, 2001, p. 230. From: Volk, W.A., Benjamin, D.C., Kadner, R.J., et al. *Essentials of Microbiology,* 4th ed. Philadelphia: J.B. Lippincott Company, 1991. Figure 32-2, p. 296.

Chapter 11

(left) *Neisseria gonorrhoeae;* modified from Figure 14.3. Strohl, W.A., et al. *Lippincott's Illustrated Reviews: Microbiology*. R.A. Harvey and P.A. Champe, eds. Philadelphia: Lippincott Williams & Wilkins, 2001, p. 166, with permission from Bottone, E.J., Girolami, R., and Stamm, J.M. *Schneierson's Atlas of Diagnostic Microbiology,* 9th ed. North Chicago: Abbott Laboratories, 1984. p. 37.

(right) *Moraxella catarrhalis;* modified from Figure 14.17. Strohl, W.A., et al. *Lippincott's Illustrated Reviews: Microbiology*. R.A. Harvey and P.A. Champe, eds. Philadelphia: Lippincott Williams & Wilkins, 2001, p. 173, with permission from Hart, T. and Shears, P., *Color Atlas of Medical Microbiology*. St. Louis: Mosby-Wolfe, 1996.

Chapter 12

(left) *Pseudomonas;* modified from Figure 16.15. Strohl, W.A., et al. *Lippincott's Illustrated Reviews: Microbiology*. R.A. Harvey and P.A. Champe, eds. Philadelphia: Lippincott Williams & Wilkins, 2001, p. 198.

(left middle) *Legionella;* modified from Figure 16.12. Strohl, W.A., et al. *Lippincott's Illustrated Reviews: Microbiology*. R.A. Harvey and P.A. Champe, eds. Philadelphia: Lippincott Williams & Wilkins, 2001, p. 196, with permission from Leboffe, M.J. and Pierce, B.E. *A Photographic Atlas for the Microbiology Laboratory*. Morton Publishing Company, 1999.

(middle) *Bordetella;* modified from Figure 16.7. Strohl, W.A., et al. *Lippincott's Illustrated Reviews: Microbiology*. R.A. Harvey and P.A. Champe, eds. Philadelphia: Lippincott Williams & Wilkins, 2001, p. 194, with permission from Leboffe, M.J. and Pierce, B.E. *A Photographic Atlas for the Microbiology Laboratory*. Morton Publishing Company, 1999.

(right middle) *Francisella;* modified from Figure 16.22. Strohl, W.A., et al. *Lippincott's Illustrated Reviews: Microbiology*. R.A. Harvey and P.A. Champe, eds. Philadelphia: Lippincott Williams & Wilkins, 2001, p. 202. From: Volk, W.A., Gebhardt, B.M., Hammarskjold, M., et al. *Essentials of Microbiology,* 5th ed. Philadelphia: Lippincott-Raven, 1996. Figure 221, p. 150, with permission from NIBSC/Science Photo Library/Photo Researcher, Inc.

(right) *Brucella;* modified from Figure 16.19. Strohl, W.A., et al. *Lippincott's Illustrated Reviews: Microbiology*. R.A. Harvey and P.A. Champe, eds. Philadelphia: Lippincott Williams & Wilkins, 2001, p. 201.

Chapter 13

(left) *Campylobacter;* modified from Figure 15.11. Strohl, W.A., et al. *Lippincott's Illustrated Reviews: Microbiology*. R.A. Harvey and P.A. Champe, eds. Philadelphia: Lippincott Williams & Wilkins, 2001, p. 181. From: Koneman, E.W., Allen, S.D., Janda, W. M., et al. *Color Atlas and Textbook of Diagnostic Microbiology,* 5th ed. Philadelphia: J.B. Lippincott Company, 1997. Plate 6-1A.

(right) *Helicobacter;* modified from Figure 15.21. Strohl, W.A., et al. *Lippincott's Illustrated Reviews: Microbiology*. R.A. Harvey and P.A. Champe, eds. Philadelphia: Lippincott Williams & Wilkins, 2001, p. 187.

Chapter 14

(left) *Escherichia coli;* modified from Figure 15.2. Strohl, W.A., et al. *Lippincott's Illustrated Reviews: Microbiology*. R.A. Harvey and P.A. Champe, eds. Philadelphia: Lippincott Williams & Wilkins, 2001, p. 176.

(middle) *Salmonella;* modified from Figure 15.7, Strohl, W.A., et al. *Lippincott's Illustrated Reviews: Microbiology*. R.A. Harvey and P.A. Champe, eds. Philadelphia: Lippincott Williams & Wilkins, 2001, p. 179. From: Volk, W.A., Benjamin, D.C., Kadner, R.J. et al. *Essentials of Microbiology,* 4th ed. Philadelphia: J.B. Lippincott Company, 1991. Figure 16-12, p. 250.

(right) *Yersinia pestis;* modified from Figure 16.27, Strohl, W.A., et al. *Lippincott's Illustrated Reviews: Microbiology*. R.A. Harvey and P.A. Champe, eds. Philadelphia: Lippincott Williams & Wilkins, 2001, p. 205, with permission from Elsevier.

Chapter 15

(left) *Vibrio;* modified from Figure 15.17. Strohl, W.A., et al. *Lippincott's Illustrated Reviews: Microbiology*. R.A. Harvey and P.A. Champe, eds. Philadelphia: Lippincott Williams & Wilkins, 2001, p. 185. From: Volk, W.A., Gebhardt, B.M., Hammarskjold, M., et al. *Essentials of Microbiology,* 5th ed. Philadelphia: Lippincott-Raven, 1996. Figure 26-5, p. 373.

(middle) *Haemophilus;* modified from Figure 16.4. Strohl, W.A., et al. *Lippincott's Illustrated Reviews: Microbiology*. R.A. Harvey and P.A. Champe, eds. Philadelphia: Lippincott Williams & Wilkins, 2001, p. 193.

(right) *Pasteurella;* modified from Figure 16.25. Strohl, W.A., et al. *Lippincott's Illustrated Reviews: Microbiology*. R.A. Harvey and P.A. Champe, eds. Philadelphia: Lippincott Williams & Wilkins, 2001, p. 204, with permission from Elsevier.

Chapter 16

(left) *Fusobacterium;* modified from Figure 17.17. Strohl, W.A., et al. *Lippincott's Illustrated Reviews: Microbiology*. R.A. Harvey and P.A. Champe, eds. Philadelphia: Lippincott Williams & Wilkins,

2001, p. 219. From Finegold, S.M., Baron, E.J., and Wexler, H.M. *A Clinical Guide to Anaerobic Infections*, 1992.

Chapter 17

(left) *Treponema;* modified from Figure 18.3. Strohl, W.A., et al. *Lippincott's Illustrated Reviews: Microbiology*. R.A. Harvey and P.A. Champe, eds. Philadelphia: Lippincott Williams & Wilkins, 2001, p. 222, with permission from Elsevier.

(middle) *Borrelia;* modified from Figure 18.8. Strohl, W.A., et al. *Lippincott's Illustrated Reviews: Microbiology*. R.A. Harvey and P.A. Champe, eds. Philadelphia: Lippincott Williams & Wilkins, 2001, p. 224, with permission from American Society Microbiology (*Microbelibrary.org*). Jeffrey Nelson, Rush University.

(right) *Leptospira;* modified from Figure 18.15. Strohl, W.A., et al. *Lippincott's Illustrated Reviews: Microbiology*. R.A. Harvey and P.A. Champe, eds. Philadelphia: Lippincott Williams & Wilkins, 2001, p. 227. From: Volk, W.A., Benjamin, D.C., Kadner, R.J. et al. *Essentials of Microbiology*, 4th ed. Philadelphia: J.B. Lippincott Company, 1991. Figure 31-5, p. 490.

Chapter 18

Rickettsia; modified from Figure 22.2. Strohl, W.A., et al. *Lippincott's Illustrated Reviews: Microbiology*. R.A. Harvey and P.A. Champe, eds. Philadelphia: Lippincott Williams & Wilkins, 2001, p. 260. From: Volk, W.A., Gebhardt, B.M., Hammarskjold, M., and Kadner, R.J. *Essentials of Microbiology*, 5th ed. Philadelphia: Lippincott-Raven, 1996. Figure 34-1, p. 459.

Chapter 19

Chlamydia; modified from Figure 20.2. Strohl, W.A., et al. *Lippincott's Illustrated Reviews: Microbiology*. R.A. Harvey and P.A. Champe, eds. Philadelphia: Lippincott Williams & Wilkins, 2001, p. 239. Cutlip, R.C. Ames: National Animal Disease Center. United States Department of Agriculture. Agriculture Research Service.

Chapter 20

(left) Naked virus model; modified from Figure 22.2. Strohl, A.W., et al. *Lippincott's Illustrated Reviews: Microbiology*. R.A. Harvey and P.A. Champe, eds. Philadelphia: Lippincott Williams & Wilkins, 2001, p. 260.

(middle) Enveloped icosahedral virus model; modified from Figure 26.1. Strohl, A.W., et al. *Lippincott's Illustrated Reviews: Microbiology*. R.A. Harvey and P.A. Champe, eds. Philadelphia: Lippincott Williams & Wilkins, 2001, p. 295.

(right) Enveloped helical virus model, from Figure 26.6. Strohl, A.W., et al. *Lippincott's Illustrated Reviews: Microbiology*. R.A. Harvey and P.A. Champe, eds. Philadelphia: Lippincott Williams & Wilkins, 2001, p. 297.

Chapter 21

(top) Adenovirus; modified from Figure 27.8A. Strohl, A.W., et al. *Lippincott's Illustrated Reviews: Microbiology*. R.A. Harvey and P.A. Champe, eds. Philadelphia: Lippincott Williams & Wilkins, 2001, p. 312.

(bottom) Herpes virus; modified from Figure 28.2B. Strohl, A.W., et al. *Lippincott's Illustrated Reviews: Microbiology*. R.A. Harvey and P.A. Champe, eds. Philadelphia: Lippincott Williams & Wilkins, 2001, p. 318. From: Volk, W.A., Gebhardt, B.M., Hammarskjold, M. et al. *Essentials of Microbiology*, 5th ed. Philadelphia: Lippincott-Raven, 1996. Figure 38-4A, p. 522.

Chapter 22

(left) Poliovirus; modified from Figure 30.2. Strohl, A.W., et al. *Lippincott's Illustrated Reviews: Microbiology*. R.A. Harvey and P.A. Champe, eds. Philadelphia: Lippincott Williams & Wilkins, 2001, p. 350. From: Volk, W.A., Benjamin, D.C., Kadner, R.J., et al. *Essentials of Microbiology*, 4th ed. Philadelphia: J.B. Lippincott Company, 1991. Figure 46-1, p. 608.

(right) HIV; modified from Figure 31.2. Strohl, A.W., et al. *Lippincott's Illustrated Reviews: Microbiology*. R.A. Harvey and P.A. Champe, eds. Philadelphia: Lippincott Williams & Wilkins, 2001, p. 360.

Chapter 23

(left, middle) Rabies from Figure 32.2A. Strohl, A.W., et al. *Lippincott's Illustrated Reviews: Microbiology*. R.A. Harvey and P.A. Champe, eds. Philadelphia: Lippincott Williams & Wilkins, 2001, p. 380. From: Fields, B.N., Knipe, D.M. and Howley, P.M. *Fields Virology*, 3rd ed. Philadelphia: Lippincott, Williams & Wilkins, 1996.

(right) Ebola; modified from Figure 32.17. Strohl A.W., et al. *Lippincott's Illustrated* Reviews: Microbiology. R.A. Harvey and P.A. Champe, eds. Philadelphia: Lippincott Williams & Wilkins, 2001, p. 390.

Chapter 24

Rotavirus electron micrograph and diagram; modified from Figure 22.2. Strohl, A.W., et al. *Lippincott's Illustrated Reviews: Microbiology*. R.A. Harvey and P.A. Champe, eds. Philadelphia: Lippincott Williams & Wilkins, 2001, p. 393, with permission from Kapikian, A.Z., Kim, H.W., and Wyatt, R.G. *Science*. 1974;185:1049–1053.

Chapter 25

Genetic reassortment in Influenza A virus; modified from Figure 32.14. Strohl, A.W., et al. *Lippincott's Illustrated Reviews: Microbiology*. R.A. Harvey and P.A. Champe, eds. Philadelphia: Lippincott Williams & Wilkins, 2001, p. 288.

Chapter 26

(left) *Histoplasma* yeast and (right) *Histoplasma* septate hyphae; modified from Figure 22.2. Strohl, A.W., et al. *Lippincott's Illustrated Reviews: Microbiology*. R.A. Harvey and P.A. Champe, eds. Philadelphia: Lippincott Williams & Wilkins, 2001, p. 260. Courtesy of Laurel Krewson.

Chapter 27

(left) Hyphae from skin scraping in KOH; modified from Figure 22-2. Schaechter, M., Engleberg, N.C., Eisenstein, B.I., et al. *Mechanisms of Microbial Disease*, 3rd ed. Philadelphia: Lippincott Willliams & Wilkins, 1998, p. 260, with permission from Elsevier.

(right) *Candida* (electron micrograph); modified from Figure 45-4 (top). Schaechter, M., Engleberg, N.C., Eisenstein, B.I., et al. *Mechanisms of Microbial Disease,* 3rd ed. Philadelphia: Lippincott Willliams & Wilkins, 1998, p. 421.

Chapter 28

(left) *Histoplasma* (yeast); modified from Figure 23.8, Strohl, A.W., et al. *Lippincott's Illustrated Reviews: Microbiology*. R.A. Harvey and P.A. Champe, eds. Philadelphia: Lippincott Williams & Wilkins, 2001, p. 272. From: Rubin, E. and Farber, J.L. *Pathology,* 2nd ed. Philadelphia: J.B. Lippincott Company, 1994. Figure 9-57, p. 416.

(middle) *Coccidioides;* modified from Figure 23.8. Strohl, A.W., et al. *Lippincott's Illustrated Reviews: Microbiology*. R.A. Harvey and P.A. Champe, eds. Philadelphia: Lippincott Williams & Wilkins, 2001, p. 272. From: Rubin, E. and Farber, J.L. *Pathology,* 2nd ed. Philadelphia: J.B. Lippincott Company, 1994. Figure 9.58, p. 417.

(right) *Blastomyces;* modified from Figure 23.8. Strohl, A.W., et al. *Lippincott's Illustrated Reviews: Microbiology*. R.A. Harvey and P.A. Champe, eds. Philadelphia: Lippincott Williams & Wilkins, 2001, p. 272. From: Koneman, E.W. and Roberts, G.D. *Practical Laboratory Mycology,* Baltimore: Williams & Wilkins, 1985.

Chapter 29

(left) *Aspergillus;* modified from Figure 47-1. Schaechter, M., Engleberg, N.C., Eisenstein, B.I., et al. *Mechanisms of Microbial Disease,* 3rd ed. Philadelphia: Lippincott Willliams & Wilkins, 1998, p. 432.

(right) *Cryptococcus;* modified from Figure 47.4. Schaechter, M., Engleberg, N.C., Eisenstein, B.I., et al. *Mechanisms of Microbial Disease,* 3rd ed. Philadelphia: Lippincott Willliams & Wilkins, 1998, p. 434.

Chapter 30

(left) *Giardia;* modified from Figure 24.4. Strohl, A.W., et al. *Lippincott's Illustrated Reviews: Microbiology,* R.A. Harvey and P.A. Champe, eds. Philadelphia: Lippincott Williams & Wilkins, 2001, p. 281. From: Gillies, R.R. and Dodds, T.C., *Bacteriology Illustrated,* 3rd ed. Philadelphia: Williams & Wilkins, 1973. p. 194.

(right) Schistosomes; modified from Figure 25.4. Strohl, A.W., et al. *Lippincott's Illustrated Reviews: Microbiology*. R.A. Harvey and P.A. Champe, eds. Philadelphia: Lippincott Williams & Wilkins, 2001, p. 291.

Chapter 31

All figures; modified from Table 6-1. Johnson, A.G., Ziegler, R.J., Lukasewycz, O.A., et al. *Board Review Series: Microbiology and Immunology,* 4th ed. Philadelphia: Lippincott Williams & Wilkins, 2002, p. 201.

Chapter 32

All three panels; modified from Figure 287-1. Gorbach, S.L., Bartlett, J.G., and Blacklow, N.R., eds. *Infectious Diseases,* 3rd ed. Philadelphia: Lippincott Willliams & Wilkins, 2004, p. 2379.

Chapter 33

All *Taenia saginata*; modified from Figure 288-5. Gorbach, S.L., Bartlett, J.G., and Blacklow, N.R., eds. *Infectious Diseases,* 3rd ed. Philadelphia: Lippincott Williams & Wilkins, 2004, p. 2387.

Chapter 34

(left) *Enterobius* egg; modified from Figure 285-7. Gorbach, S.L., Bartlett, J.G., and Blacklow, N.R., eds. *Infectious Diseases,* 3rd ed. Philadelphia: Lippincott Williams & Wilkins, 2004, p. 2362.

(right) *Ascaris* egg; modified from Figure 285-4. Gorbach, S.L., Bartlett, J.G., and Blacklow, N.R. eds. *Infectious Diseases,* 3rd ed. Philadelphia: Lippincott Williams & Wilkins, 2004, p. 2359.

Index

Page numbers followed by *f* refer to illustrations; page numbers followed by *t* refer to tables.

A

Abscesses, 182
Absidia spp., 146, 178
Acanthamoeba spp., 156, 176
Actinomyces spp., 56*t*, 59, 182, 193
Adaptive immunity, 105–106
Adeno-associated viruses, 110, 110*t*
Adenoviruses, 111, 175, 177, 178, 203
Aerobes, 14, 15*t*
African sleeping sickness, 157, 202*t*
Agar, 45, 47*t*, 137, 201*t*
Albendazole, 171
Alpha-toxin, 21*t*
Alphaviruses, 119
Alveolar hydatid cyst, 167
Amebas, 155–156
Amebic dysentery, 155
Aminoglycosides, 40*t*
Anaerobes, 14, 15*t*, 83
Ancylostoma duodenale, 170, 188
Anemias, 188
Anopheles mosquitos, 158
Anthrax, 23*t*, 25, 55, 56
Antibiotics
  bacterial resistance, 40*t*, 41–43
  combination therapy, 41
  effectiveness, 39
  effects in bacteria, 11*t*
  mechanisms of action, 39, 40*t*, 41*f*
Antigenic drift, 128
Arenaviruses, 123*t*, 126
Arthritis, septic, 189, 207–208
Ascaris lumbricoides, 169, 186
Aspergillosis, 137, 144–145
Aspergillus fumigatus, 144–145, 182, 208

B

Bacillary angiomatosis, 91*t*, 92
Bacillus spp.
  B. anthracis, 23*t*, 55, 56, 194
  B. cereus, 56, 184
  characteristics, 56*t*
  structure, 13*f*
Bacitracin, 40*t*, 41*f*
Bacteria (*see also specific bacteria*)
  aerobic, 15*t*
  anaerobic, 15*t*
  characteristics, 4
  commensals, 14
  DNA molecules in, 27–28
  genera, 48–49, 48*f*, 49*f*
  genetics and reproduction
    conjugation, 29, 31*f*, 32*f*, 33*f*
    homologous recombination, 28–29
    mutations, 28
    transduction, 32, 34–38*f*, 36
    transformation, 29, 30*f*
  Gram-negative (*see* Gram-negative bacteria)
  Gram-positive (*see* Gram-positive bacteria)
  growth, 15, 16*f*
  laboratory identification, 44–47, 48*t*
  metabolism, 14
  pathogenicity
    colonization of human body, 18–19
    entry into human cells and tissues, 19
    evasion of host defense system, 19–20
    intracellular growth, 26
    survival outside human host, 17–18
    toxins, 20, 21–24*t*, 25
  poorly seen on Gram stain, 63
  structures
    cell envelope, 7, 8*f*, 9*f*
    chromosomes, 12
    endospores, 12
    flagella, 12, 12*f*
    interior, 12
    pili, 12
    protrusions, 12, 12*f*
    surface antigens, 12
Bacterial vaginosis, 191
Bacteroides, 83
Bartonella, 91*t*, 92, 187
Beef tapeworm, 166–167
Beta-lactams, 41*f*, 53
Biofilms, 19
Bladder cancer, 165, 192
Blastomyces dermatitidis, 143, 203
Blastomycosis, 143
Blood, in urine, 190
Blood flukes, 164–165
Boils, 193
Bone infections, 51, 53, 78, 189, 209
Bordetella pertussis
  characteristics, 69
  diseases caused by, 69, 178, 188
  laboratory tests, 69
  prevention of infection, 70
  specimen collection, 45
  toxin, 24*t*
  virulence, 69
Borrelia spp., 87–88
Botulinum toxin, 24*t*, 25
Botulism, 57–58
Brain abscesses, 182
Broad fish tapeworm, 167
Bronchiolitis, 124, 178
Bronchitis, 178

Brucella spp., 26, 70
Brucellosis, 70
Brugia malayi, 171
Bubonic plague, 79, 202*t*
Bullous impetigo, 50
Bunyaviruses, 123*t,* 126
Burkitt lymphoma, 114, 192
Burn patients, *Pseudomonas* infection in, 68

C

Caliciviruses, 117*t,* 118
California encephalitis virus, 126
Campylobacter spp., 71, 185, 202*t*
Cancer
  associated with parasites, 192
  associated with viruses, 192
  Burkitt lymphoma, 114, 192
  cervical, 111, 192
  liver, 107, 192
Candida, 139, 145–146, 191
Candidiasis, 145
Cardiovascular infections, 51, 54, 145, 187
Case studies, 207–210
Cat scratch fever, 91*t,* 92
Catalase, 48*t*
Cell envelope, of bacteria, 7, 9*f,* 10*f,* 11*t*
Cephalosporins, 40*t*
Cervical carcinoma, 111, 192
Cervical intraepithelial neoplasia, 209
Cestodes
  characteristics, 153, 166
  Diphyllobothrium latum, 167, 188
  intestinal, 166–167
  Taenia saginata, 166–167
  Taenia solium, 166–167
  tissue, 167
Chagas disease, 157, 187, 202*t*
Chancroid, 82
Chickenpox, 113
Chinese liver fluke, 163
Chlamydia spp.
  C. pneumoniae, 95
  C. psittaci, 95, 179
  C. trachomatis, 93–95, 175, 179, 190
  characteristics, 93
  drug resistance, 43
  life cycle, 93, 94*f*
  virulence, 93
Chloramphenicol, 40*t,* 41*f*
Cholera, 23*t,* 25, 80, 209
Chronic mucocutaneous candidiasis, 145
Clindamycin, 40*t,* 41*f*
Clonorchis sinensis, 163
Clostridium spp.
  C. botulinum, 24*t,* 57–58, 183
  C. difficile, 58, 186, 209
  C. perfringens
    diseases caused by, 57, 194
    epidemiology, 57
    toxins, 21*t,* 22*t,* 25
  C. tetani
    diseases caused by, 56–57, 183, 194
    epidemiology, 56
    toxin, 24*t*
  characteristics, 56*t*
  structure, 13*f*
Coagulase, 48*t,* 200
Coccidioides immitis, 142, 203
Coccidioidomycosis, 142
Colorado tick fever virus, 127
Commensals, 14
Common cold, 118, 120, 177
Complementation, 129
Conjugation, in bacterial gene transfer, 29, 31*f,* 32*f,* 33*f*
Conjunctivitis, 94, 111, 175
Coronaviruses, 117*t,* 120
Corynebacterium spp.
  C. diphtheriae
    colonization patterns, 18
    diseases caused by, 59, 177, 209
    epidemiology, 59
    toxin, 22*t,* 25
    virulence, 59
  characteristics, 56*t,* 58
Coxiella burnetii, 91–92, 91*t*
Coxsackievirus, 118, 177
Creutzfeldt-Jakob disease, 183
Croup, 124, 178
Cryptococcus neoformans, 146, 181, 202
Cryptosporidium parvum, 158, 185
Cultures, 45, 47, 47*t*
Cutaneous anthrax, 56
Cutaneous larva migrans, 171, 194
Cycloserine, 40*t,* 41g
Cystic fibrosis, *Pseudomonas* infection in, 68, 180
Cysticercosis, 167
Cystitis, 73, 190
Cytomegalovirus, 114
Cytotoxic T cells, 106

D

Defective interfering particles, 128
Dengue virus, 119
Dermatophytes, 138–139, 193
Dermatophytid reaction, 139
Diaper rash, 139
Diarrhea
  C. difficile, 209
  case studies, 184–186
  E. coli, 75, 185, 202*t,* 209
  inflammatory, 155, 185–186, 202*t,* 209
  noninflammatory, 184–185, 202*t*
  rotavirus, 127, 202*t*
Diphtheria, 22*t,* 25, 59, 177, 209
Diphyllobothrium latum, 167, 188
DNA
  in bacteria, 4*t,* 27–28
  in eukaryotic cells, 4*t*
DNA viruses
  adenoviruses, 111
  characteristics, 109, 110*t*
  hepadnavirus, 111–112, 112*f*
  herpesviruses (*see* Herpesviruses)
  papovaviruses, 110*t,* 111, 183
  parvoviruses, 110, 110*t,* 195
  replication, 102*t,* 109
Double-stranded RNA viruses, 127
Dracunculus, 171
Drug resistance (*see also specific bacteria*)
  acquisition, 42–43
  bacteria with, 43
  genetics, 42
  mechanisms, 40*t,* 41–42
  Plasmodium falciparum, 161
Dysentery, 155, 186

E

Ear infections, 176
Ebola virus, 125
Echinococcus granulosis, 167
Echinococcus multilocularis, 167
Echoviruses, 118
Ecthyma gangrenosum, 68
Ehrlichia, 91*t,* 92
Ehrlichiosis, 91*t*
Elephantiasis, 171
Encephalitis, 119, 126, 182
Endocarditis, 51, 54, 145, 187

Endospores, 12
Endotoxins, 20, 21*t*
Entamoeba spp.
  E. coli, 155
  E. dispar, 155
  E. histolytica, 155–156, 186, 203, 209
Enterobacteriaceae
  characteristics, 73
  drug resistance, 43
  Escherichia coli (*see Escherichia coli*)
  genera, 73, 74*t*, 79
  virulence factors, 73, 74*f*
Enterobius vermicularis, 169–170
Enterococcus spp.
  characteristics, 54
  diseases caused by, 54, 187
  drug resistance, 43, 54
Enteroviruses, 118, 181
Envelope surface antigens, 12
Enzyme tests, 47, 48*t*
Epidemic typhus, 91*t*
Epidermophyton, 138–139
Epiglottitis, 81, 178
Episomes, 27
Epstein-Barr virus, 114, 177, 192, 195, 210
Equine encephalitis virus, 119
Erysipelas, 53
Erythema infectiosum, 110
Escherichia coli
  characteristics, 73
  colonization mechanisms, 18
  diseases caused by, 73–75, 190, 202*t*
  enterohemorrhagic, 75, 129–130, 185
  enteroinvasive, 75
  enteropathogenic, 75, 185
  enterotoxic, 75, 184, 209
  laboratory tests, 75
  survival outside human host, 17
  toxins, 23*t*, 129–130
Ethambutol, 41*f*
Eukaryotes, 3, 4*t*
Exanthem subitum, 114, 195
Exotoxin A, 22*t*, 25
Exotoxins, 20, 25
Eye infections, 175–176
Eye worm, 171

F

Farmer's lung, 137, 144
Fasciola hepatica, 163
Fasciolopsis buski, 163
Fatal familial insomnia, 183
Favus, 139
Fertility factor, 29, 31*f*
Fifth disease, 110
Filarial worms, 170–171
Filoviruses, 123*t*, 125
Flagella, 12, 12*f*
Flagellates
  Giardia lamblia, 156, 185, 202*t*, 203, 209
  Leishmania, 158, 202*t*
  Trichomonas vaginalis, 156–157, 191
  Trypanosoma brucei, 157
  Trypanosoma cruzi, 157–158, 175, 187, 208
Flaviviruses, 117*t*, 119
Flukes
  blood, 164–165
  characteristics, 152, 163, 164*t*
  intestinal, 163
  liver, 163
  lung, 164
  treatment, 165
Fluoroquinolones, 40*t*, 41*f*
Food poisoning
  B. cereus, 56
  C. botulinum, 57
  C. perfringens, 57
  case studies, 184–185
  S. aureus, 50
Francisella, 70
Fungi (*see also specific fungi*)
  Absidia, 146
  Aspergillus fumigatus, 144–145, 182, 208
  Blastomyces dermatitidis, 143
  Candida, 139, 145–146, 191
  characteristics, 4, 135, 203
  Coccidioides immitis, 142, 203
  Cryptococcus neoformans, 146, 181, 202
  dermatophytes, 138–139
  diseases caused by, 136–137
  Histoplasma capsulatum, 141–142, 202*t*, 203
  laboratory identification, 137
  Malassezia furfur, 138
  Mucor, 146
  opportunistic, 144–147
  Pneumocystis jiroveci, 147
  Rhizopus, 146
  Sporothrix schenckii, 139–140, 193, 203
  sporulation, 136
  structures, 135–136
  systemic pathogens, 141–143
Fusobacterium, 84

G

Gardnerella vaginalis, 191
Gas gangrene, 57
GAS (group A streptococci) (*see Streptococcus* spp., *S. pyogenes*)
Gastrointestinal infections, 118, 184–186 (*see also* Diarrhea; Food poisoning)
Gene probes, 47
Gene transfer, bacteria
  conjugation, 29, 31*f*, 32*f*, 33*f*
  transduction, 32, 34–38*f*, 36
  transformation, 29, 30*f*
Genetic drift, 128
Genetic shift, 129, 130*f*
Genital infections, 190–191 (*see also* Sexually transmitted diseases)
Giardia lamblia, 156, 185, 202*t*, 203, 209
Glomerulonephritis, acute, 53
Gonococcus (*see Neisseria* spp., *N. gonorrhoeae*)
Gonorrhea, 66, 191
Gram-negative bacteria
  Bacteroides, 83
  Bordetella (*see Bordetella pertussis*)
  Brucella spp. (*see Brucella* spp.)
  Campylobacter spp., 71, 185, 202*t*
  cell envelope, 7, 9*f*, 10*f*, 11*t*
  Francisella, 70
  Fusobacterium, 84
  genera, 49*f*
  Haemophilus spp. (*see Haemophilus* spp.)
  Helicobacter pylori, 71–72
  Klebsiella spp., 76
  Legionella (*see Legionella*)
  Moraxella spp., 66
  Neisseria spp. (*see Neisseria* spp.)
  Pasteurella, 82
  Porphyromonas, 84
  Prevotella, 84
  Proteus spp., 78
  Pseudomonas spp. (*see Pseudomonas* spp.)
  Salmonella spp. (*see Salmonella* spp.)
  Shigella spp. (*see Shigella* spp.)
  toxins, 20, 21*t*
  Vibrio spp. (*see Vibrio* spp.)
  Yersinia spp. (*see Yersinia* spp.)
Gram-positive bacteria
  Actinomyces, 59
  Bacillus spp. (*see Bacillus* spp.)
  cell envelope, 7, 8*f*, 10*f*, 11*t*

Gram-positive bacteria (*Cont.*)
- *Clostridium* spp. (*see Clostridium* spp.)
- *Corynebacterium* spp. (*see Corynebacterium* spp.)
- differentiation, 56*t*
- genera, 48*f*, 56*t*
- *Listeria* spp. (*see Listeria* spp.)
- *Mycobacteria* (*see Mycobacterium* spp.)
- *Nocardia*, 56*t*, 59, 182
- staphylococci (*see Staphylococcus* spp.)
- streptococci (*see Streptococcus* spp.)
- toxins, 21*t*

Gram stain, 10*t*, 47
Granulomatous amebic meningoencephalitis, 156
Group A streptococci (GAS) (*see Streptococcus* spp., *S. pyogenes*)
Group B streptococci (GBS), 53–54

H

Haemophilus spp.
- characteristics, 81
- *H. ducreyi*, 82, 191
- *H. influenzae*
  - conjunctivitis, 175
  - drug resistance, 43
  - epidemiology, 81
  - epiglottitis, 81, 178
  - laboratory tests, 81
  - meningitis, 81, 181
  - otitis media, 81, 176
  - pneumonia, 179

Hand-foot-and-mouth disease, 118
Hantavirus, 126
Helicobacter pylori, 71–72
Hemolytic uremic syndrome, 75
Hemorrhagic fever, 125, 126
Hepadnavirus, 111–112, 112*f*
Hepatitis, 107, 108*t*, 186, 202*t*
Hepatitis B, 108*t*, 111–112, 112*f*
Hepatitis C, 107, 108*t*, 117*t*, 119
Hepatitis D, 129
Hepatitis E, 107, 108*t*, 117*t*, 118
Hepatocellular carcinoma, 107, 192
Herpes zoster, 114
Herpesviruses
- characteristics, 113
- cytomegalovirus, 114
- Epstein-Barr virus, 114
- herpes simplex viruses, 113, 182, 183, 190
- poxviruses, 114
- varicella-zoster virus, 113–114

Heterotrophs, 14
Histoplasma capsulatum, 141–142, 202*t*, 203
Histoplasmosis, 142, 202*t*
Homologous recombination, 28–29, 28*f*
Hookworms, 170, 188
Hosts, 151
Human herpesvirus 6, 114, 195 (*see also* Herpesviruses)
Human immunodeficiency virus, 117*t*, 120–121, 128
Human papillomaviruses, 111, 190, 209
Human T cell leukemia virus, 192
Hydatid cyst disease, 167
Hyphae, 135

I

Immune response, 104–106
Impetigo, 53, 193
Infections
- blood cell count changes in, 188
- bone, 51, 53, 78, 189
- cardiovascular, 51, 54, 145, 187
- case studies, 207–210
- causative agents, reservoirs, and transmission modes, 202*t*
- ear, 176
- eye, 175–176
- gastrointestinal, 184–186 (*see also* Diarrhea; Food poisoning)
- genital tract, 190–191 (*see also* Sexually transmitted diseases)
- nervous system, 181–183 (*see also* Encephalitis; Meningitis)
- respiratory tract, 177–180 (*see also* Pharyngitis; Pneumonia)
- septicemia, 188
- skin and subcutaneous, 138–139, 193–194
- urinary tract (*see* Urinary tract infection)

Infectious mononucleosis, 114, 177, 210
Influenza, 125–126, 128, 178
Innate immunity, 104–105
Interferons, 105
Isoniazid, 40*t*, 41*f*
Ixodes, 88

J

JC virus, 183

K

Kala-azar, 158
Kaposi sarcoma, 114
Keratitis, 156
Klebsiella spp., 76
Koplik spots, 124
Kuru, 183

L

Labile toxin, 23*t*, 25
Laboratory tests
- specificity vs. sensitivity, 45, 46*f*
- specimen handling, 45
- types, 44–45, 200–201, 201*t*

Lassa fever, 126
Legionella
- characteristics, 68
- diseases caused by, 68, 179, 180, 202*t*
- laboratory tests, 68–69
- prevention of infection, 69
- survival outside human host, 17
- treatment, 69

Leishmania spp., 158, 202*t*
Leprosy, 61, 61*t*, 202*t*
Leptospira, 88–89, 182, 202*t*
Linezolid, 40*t*
Listeria spp.
- characteristics, 56*t*, 58
- *L. monocytogenes*
  - diseases caused by, 58, 181
  - epidemiology, 58
  - evasion of host defenses, 20
  - intracellular growth, 26
  - survival outside human host, 17
  - toxin, 21*t*, 25
  - virulence, 58

Listeriolysin O, 21*t*, 25
Liver carcinoma, 107, 192
Liver flukes, 163
Loa loa, 171
Lung fluke, 164
Lyme disease, 88
Lymphocytic choriomeningitis, 126
Lymphogranuloma venereum, 95
Lysogeny, 36*f*, 129

M

M cells, 19, 26
MAC (membrane activation complex), 20
Macrolides, 40*t*, 41*f*
Malaria, 159–160, 160*t*, 202*t*
Malassezia furfur, 138

Marburg virus, 125
Measles, 124, 183, 195
Membrane activation complex (MAC), 20
Meningitis
case studies, 181–182, 207, 208
cryptococcal, 202*t*
E. coli, 75
H. influenzae, 81
L. monocytogenes, 58
N. meningitidis, 65, 195, 208
S. agalactiae, 207
S. pneumonia, 52
viral, 118, 181
Meningococcus (*see Neisseria* spp., *N. meningitidis*)
Meningoencephalitis, 156, 182
Metronidazole, 40*t*, 41*f*
Microsporum, 138–139
Mobilization, 42
Molluscum contagiosum, 115, 194
Moraxella spp., 66, 176
Mucor, 146, 178
Mumps, 124
Mycetismus, 137
Mycobacterium spp.
characteristics, 56*t*, 60, 203, 208
intracellular growth, 26
M. avium-intracellulare, 61
M. leprae, 61, 61*t*, 182
M. marinum, 61
M. tuberculosis
characteristics, 60, 208
diagnostic tools, 60–61
diseases caused by, 60, 182, 202*t*, 208
drug resistance, 43
evasion of host defenses, 20
Mycoplasma spp.
characteristics, 62
drug resistance, 43
M. hominis, 62
M. pneumoniae, 20, 62, 179
Mycotoxicosis, 137
Myonecrosis, 57
Myopericarditis, 118, 187

N

Naegleria spp., 156, 182
Natural killer (NK) cells, 104–105
Necator americanus, 170, 188
Necrotizing fasciitis, 53, 194
Negative RNA viruses
bunyaviruses, 123*t*, 126
characteristics, 122, 123*t*
filoviruses, 123*t*, 125
orthomyxoviruses, 123*t*, 125–126
paramyxoviruses, 123*t*, 124
replication, 102*t*, 122
rhabdoviruses, 123*t*, 124–125
Negri bodies, 125
Neisseria spp.
characteristics, 64
N. gonorrhoeae
characteristics, 65
conjunctivitis, 66, 175
drug resistance, 42, 43, 66
gonorrhea, 66, 191
laboratory tests, 66
polyarticular arthritis, 189, 207–208
prevention of infection, 66
treatment, 66
virulence, 65–66
N. meningitidis
characteristics, 64, 65*f*, 208
diseases caused by, 65, 181, 195, 208
entry through mucosal surfaces, 19
epidemiology, 64
laboratory tests, 65
prevention of infection, 65
virulence, 64
specimen handling, 45
Nematodes
Ancylostoma duodenale, 170, 188
animal roundworms, 171
Ascaris lumbricoides, 169, 186
Brugia malayi, 171
characteristics, 152, 168–169
Dracunculus, 171
Enterobius vermicularis, 169–170
intestinal, 169–170
Loa loa, 171
Necator americanus, 170, 188
Onchocerca volvulus, 171
Strongyloides stercoralis, 170
tissue, 170–171
treatment, 171
Trichinella spiralis, 170, 175
Trichuris trichiura, 170, 188
Wuchereria bancrofti, 171
Nervous system infections, 181–183 (*see also* Encephalitis; Meningitis)
Nocardia spp., 56*t*, 59, 182
Norovirus, 118, 202*t*
Norwalk agent gastroenteritis, 118, 184

O

Obligate intracellular pathogens, 26
Onchocerca volvulus, 171
Opportunistic infections, 121, 144–147
Orbiviruses, 127
Orthomyxoviruses, 123*t*, 125–126
Orthoreoviruses, 127
Osteomyelitis, 51, 53, 78, 189, 209
Otitis media, 52, 62, 81, 176
Oxidase, 48*t*

P

Papovaviruses, 110*t*, 111, 183
Paragonimus westermani, 164
Parainfluenza virus, 124, 178
Paramyxovirus, 123*t*, 124
Parasites
cancers associated with, 192
cestodes (*see* Cestodes)
characteristics, 4, 151
diseases caused by, 152
flukes
blood, 164–165
characteristics, 152, 163, 164*t*
intestinal, 163
liver, 163
lung, 164
treatment, 165
hosts, 151
laboratory tests, 152, 153*f*
nematodes (*see* Nematodes)
protozoa
amebas, 155–156
characteristics, 152
flagellates (*see* Flagellates)
sporozoa (*see* Sporozoa)
vectors, 152
Paresthesia, 182
Parvoviruses, 110, 110*t*, 195
Pasteurella, 82
Pathogenicity islands (PAIs), 27
Penicillins, 40*t*
Peptidoglycan
in Gram-negative bacteria, 7, 10*f*, 11*t*
in Gram-positive bacteria, 7, 10*f*, 11*t*
Peptostreptococci, 54

Pericarditis, 187
Pertussis, 24*t*, 25, 69, 178
Phage DNA, 28
Pharyngitis
  adenovirus, 111, 177
  case study, 208
  in diphtheria, 59
  in infectious mononucleosis, 114, 210
  streptococcal, 53, 177
Phenotypic masking, 131
Phenotypic mixing, 131
Picornaviruses, 116, 117*t*, 118
Pili, 12
Pinworm, 169–170
Plague, 79, 202*t*
Plasmid DNA, 27–28
Plasmodium spp.
  cancer associated with, 192
  characteristics, 158, 160*t*
  epidemiology, 158–159
  laboratory tests, 160
  life cycle, 159*f*
  P. falciparum, 159, 160*t*
  P. malariae, 159, 160*t*
  P. ovale, 158, 160*t*
  P. vivax, 158, 160*t*
  prophylaxis and treatment, 160–161
Pneumococcus (*see Streptococcus* spp., *S. pneumoniae*)
Pneumocystis jiroveci (*P. carinii*), 147
Pneumolysin O, 52
Pneumonia
  anthrax, 56
  in aspergillosis, 145
  C. pneumoniae, 95
  C. psittaci, 95
  C. trachomatis, 94, 179
  case studies, 178–180, 207
  fungal, 142–143
  K. pneumoniae, 76
  M. pneumoniae, 62
  P. aeruginosa, 68
  Pneumocystis, 147
  S. aureus, 51
  S. pneumoniae, 52
Poliovirus, 118
Polymyxins, 40*t*, 41*f*
Polyomaviruses, 111
Pork roundworm, 170
Pork tapeworm, 166–167
Porphyromonas, 84
Positive RNA viruses
  caliciviruses, 117*t*, 118
  characteristics, 116, 117*t*
  coronaviruses, 117*t*, 120
  definition, 116
  flaviviruses, 117*t*, 119
  picornaviruses, 116, 117*t*, 118
  replication, 102*t*
  retroviruses, 102*t*, 117*t*, 120–121
  togaviruses, 117*t*, 119–120
Poxviruses, 114
Praziquantel, 165
Prevotella, 84
Primary amebic meningoencephalitis, 156
Prions, 3–4, 183
Prokaryotes, 3, 4*t*
Proteus spp., 78
Protozoa
  amebas, 155–156
  characteristics, 152
  flagellates (*see* Flagellates)
  sporozoa (*see* Sporozoa)
Pseudomonas spp.
  in burn patients, 68
  cellulitis, 194, 207
  characteristics, 67
  in cystic fibrosis patients, 68
  in diabetic patients, 68
  drug resistance, 43, 68
  eye infection, 176
  in immunocompetent people, 67–68
  osteomyelitis, 189
  pneumonia, 180
  survival outside human host, 18
  toxin, 22*t*
  virulence, 67
Psittacosis, 95
Purified protein derivative, 60
Pyelonephritis, 74

Q

Q fever, 91*t*, 92

R

Rabies, 124–125
Rashes, 194–195
Reoviruses, 127
Respiratory syncytial virus, 124, 178
Respiratory tract infections, 177–180 (*see also* Pharyngitis; Pneumonia)
Retroviruses, 102*t*, 117*t*, 120–121
Rhabdoviruses, 123*t*, 124–125
Rheumatic fever, 53
Rhinoviruses, 18, 177
Rhizopus, 146, 178
Rickettsia spp.
  characteristics, 90, 91*t*
  diseases caused by, 90, 91*t*, 202*t*
  drug resistance, 43
  epidemiology, 90
  intracellular growth, 26
  laboratory tests, 91
  treatment, 91
Rickettsial pox, 91*t*
Rifampin, 40*t*, 41*f*, 65
River blindness, 171
RNA viruses
  double-stranded, 127
  positive (*see* Positive RNA viruses)
  negative (*see* Negative RNA viruses)
Rocky Mountain spotted fever, 91*t*, 202*t*
Rose gardener's disease, 140
Roseola, 114
Rotaviruses, 127, 184, 202*t*
Roundworms (*see* Nematodes)
Rubella, 120
Rubeola, 124

S

Sabin vaccine, 118
Salk vaccine, 118
Salmonella spp.
  characteristics, 77, 202*t*
  in osteomyelitis, 189
  S. enterica, 77–78
  S. enteritidis, 185, 209
Scalded skin syndrome, 50
Scarlet fever, 53, 195
Schistosoma spp.
  S. haematobium, 165, 190, 192, 202*t*
  S. japonicum, 164
  S. mansoni, 164
Sensitivity, 45, 46*f*
Septicemia, 188
Sexually transmitted diseases
  case studies, 190–191
  chlamydia, 94, 175, 179, 190
  gonorrhea, 66, 191
  lymphogranuloma venereum, 95
  syphilis, 85–87, 86*f*, 87*t*, 191

Sheep liver fluke, 163
Shiga toxin, 23*t*, 25, 75, 76
Shigella spp.
  characteristics, 76
  diseases caused by, 76, 186
  intracellular growth, 26
  laboratory tests, 76–77
  S. dysenteriae, 23*t*, 183, 186
  virulence, 76
Shingles, 114, 195
Shock, 188
Sick building syndrome, 137
Skin infections, 138–139, 193–194
Smallpox, 115
South American hemorrhagic fever, 126
Specificity, 45, 46*f*
Spirochetes
  Borrelia spp., 87–88
  characteristics, 85
  Leptospira, 88–89, 182, 202*t*
  Treponema (*see Treponema pallidum*)
Sporothrix schenckii, 139–140, 193, 203
Sporotrichosis, 140
Sporozoa
  Cryptosporidium parvum, 158
  Plasmodium spp. (*see Plasmodium* spp.)
  Toxoplasma gondii, 161–162, 161*f*, 176
St. Louis encephalitis, 119
Stain reactions, 200–201
Staphylococcus spp.
  S. aureus
    bone infection, 51, 189
    cardiovascular infection, 51, 187
    case study, 208
    characteristics, 50
    drug resistance, 42, 43, 51
    epidemiology, 50
    eye infections, 175
    gastrointestinal infection, 50, 184
    laboratory identification, 51
    skin infection, 50, 193
    survival outside human host, 17
    toxic shock syndrome, 50–51
    toxins, 21*t*, 22*t*, 25
  S. epidermidis
    biofilm formation, 19
    characteristics, 51
    diseases caused by, 51, 187
  S. pyogenes, 17
  S. saprophyticus
    characteristics, 51
    diseases caused by, 51, 190
"Strep throat," 53
Streptococcus spp.
  characteristics, 51, 52*t*
  S. agalactiae, 53–54, 181, 207, 210
  S. mutans, 19, 54
  S. pneumoniae
    characteristics, 52
    colonization patterns, 18
    drug resistance, 43, 52
    epidemiology, 52
    eye infection, 175
    laboratory identification, 52
    meningitis, 52, 181
    otitis media, 52, 176
    respiratory infection, 52, 177, 179
    vaccination, 52–53
    virulence, 52
  S. pyogenes
    characteristics, 53
    diseases caused by, 53, 177, 194, 208
    laboratory identification, 53
    prevention of infection, 53
    toxins, 21*t*, 22*t*, 25
    virulence, 53
  viridans, 54, 187
Streptogramins, 40*t*
Streptolysin O, 21*t*
Strongyloides stercoralis, 170
Subacute sclerosing panencephalitis, 124, 183
Subcutaneous infections, 193–194
Sulfonamides, 40*t*, 41*f*
Swimmer's ear, 68
Swimmer's itch, 165, 193
Syphilis, 85–87, 86*f*, 87*t*, 191

T

Taenia saginata, 166–167
Taenia solium, 166–167
Tapeworms (*see* Cestodes)
Tetanus, 24*t*, 25, 56–57, 57*t*
Tetracycline, 40*t*, 41*f*
Threadworm, 170
Thrush, 139, 145
Tinea barbae, 139
Tinea capitis, 139
Tinea corporis, 139
Tinea cruris, 139
Tinea favosa, 139
Tinea pedis, 139
Tissue cestodes, 167
Tissue nematodes, 170–171
Togaviruses, 117*t*, 119–120
Toxic shock syndrome toxin 1 (TSST-1), 22*t*, 25, 50–51, 188
Toxins
  disease effects, 21–24*t*
  mechanisms of action, 21–24*t*, 199–200, 200*t*
  types, 20, 21–24*t*, 25
Toxocara, 171
Toxoplasma gondii, 161–162, 161*f*, 176
Toxoplasmosis, 162, 202*t*
Trachoma, 94–95, 175
Transduction, in bacterial gene transfer, 32, 34–38*f*, 36
Transfection, 131
Transformation, in bacterial gene transfer, 29, 30*f*
Traveler's diarrhea, 75
Trematodes (*see* Flukes)
Treponema pallidum
  characteristics, 85
  diseases caused by, 85–86, 191
  epidemiology, 85
  intracellular growth, 26
  laboratory tests, 86–87, 86*f*, 87*t*
Trichinella spiralis, 170, 175
Trichomonas vaginalis, 156–157, 191
Trichophyton, 138–139
Trichuris trichiura, 170, 188
Trimethoprim, 40*t*, 41*f*
Trypanosoma brucei, 157
Trypanosoma cruzi, 157–158, 175, 187, 208
Tuberculosis, 60–61, 202*t*
Typhoid, 77, 202*t*

U

Undulant fever, 70
Ureaplasma urealyticum, 62
Urease, 48*t*, 200
Urinary tract infection
  E. coli, 73, 190
  Klebsiella, 76
  S. saprophyticus, 51, 190

V

Vaccine
  H. influenzae, 81
  measles, 124
  mumps, 124
  N. meningitidis, 65, 181
  pneumococcal, 52–53

Vaccine (*Cont.*)
poliovirus, 118
rabies, 125
tetanus, 56, 57*t*
Vaccinia virus, 115
Vaginitis, 139, 190–191
Valley fever, 142, 202*t*
Vancomycin, 40*t*, 41*f*
Varicella-zoster virus, 113–114, 182, 195
Variola virus, 115
Vectors, 131, 152
Verotoxin, 23*t*
Vesicular stomatitis virus, 125
Vibrio spp.
characteristics, 80
V. cholerae, 80, 184, 209
V. parahaemolyticus, 81, 185
V. vulnificus, 81
Virions, 99, 100, 100*f*
Virulence factors, 199–200
Viruses (*see also specific viruses*)
cancers associated with, 192
characteristics, 4
disease patterns, 106–107, 106*f*
enveloped, 101
genetics
complementation, 128–129
genetic shift, 129
latency, 129–130
mutation, 128
phenotypic masking and mixing, 131
transfection, 131
host cell range, 101
host defenses, 104–106
latent in nerves, 182–183
naked, 100
one-step growth curve, 105*f*
replication, 101–103, 102*f*, 102*t*, 103*f*, 104*f*, 105*f*
structure, 99–101, 100*f*
as vectors, 131

W

Warts, 111
West Nile virus, 119
Whipworm, 170
White blood changes, 188
Winterbottom sign, 157
Wuchereria bancrofti, 171

Y

Yellow fever, 119
Yersinia spp.
characteristics, 78
survival outside human host, 17
Y. enterocolitica, 79
Y. pestis, 78–79, 180, 202*t*

Z

Zygomycota, 146